High Altitude
Geoecology

AAAS Selected Symposia Series

High Altitude Geoecology

Edited by Patrick J. Webber

Routledge
Taylor & Francis Group

LONDON AND NEW YORK

AAAS Selected Symposium 12

First published 1979 by Westview Press, Inc.

Published 2018 by Routledge
52 Vanderbilt Avenue, New York, NY 10017
2 Park Square, Milton Park, Abingdon, Oxon OX14 4RN

Routledge is an imprint of the Taylor & Francis Group, an informa business

Library of Congress Catalog Card Number: 78-20319

ISBN 13: 978-0-367-02118-4 (hbk)
ISBN 13: 978-0-367-17105-6 (pbk)

About the Book

This collection of papers is concerned with the ecology and occupation of mountain areas. Focusing especially on ecological problems in these areas, its theme is an outgrowth of "Man and the Biosphere," a fledgling international program sponsored by UNESCO. The book is particularly concerned with section 6 of the UNESCO program, "Impact of Human Activities on Temperate and Tropical Mountain and Tundra Ecosystems."

Each of the contributing authors is an internationally recognized authority, and each has provided a review of the state of knowledge and a discussion of special problems and areas for future research in his or her field of specialization.

Contents

List of Figures and Tables

Foreword

The *AAAS Selected Symposia Series* was begun in 1977 to provide a means for more permanently recording and more widely disseminating some of the valuable material which is discussed at the AAAS Annual National Meetings. The volumes in this *Series* are based on symposia held at the Meetings which address topics of current and continuing significance, both within and among the sciences, and in the areas in which science and technology impact on public policy. The *Series* format is designed to provide for rapid dissemination of information, so the papers are not typeset but are reproduced directly from the camera copy submitted by the authors, without copy editing. The papers are reviewed and edited by the symposia organizers who then become the editors of the various volumes. Most papers published in this *Series* are original contributions which have not been previously published, although in some cases additional papers from other sources have been added by an editor to provide a more comprehensive view of a particular topic. Symposia may be reports of new research or reviews of established work, particularly work of an interdisciplinary nature, since the AAAS Annual Meeting typically embraces the full range of the sciences and their societal implications.

WILLIAM D. CAREY
Executive Officer
American Association for
the Advancement of Science

About the Editor and Authors

Patrick J. Webber, associate professor in the Department of Biology and the Institute for Arctic and Alpine Research at the University of Colorado, is a geoecologist specializing in the structure and function of vegetation, especially in tundra areas. His recent projects have included Prudhoe Bay Vegetation Mapping and the Meade River RATE Program, and he is a coprincipal investigator with the International Biological Program's Tundra Biome Program. *He is a member of the editorial board of* Arctic and Alpine Research *and has published numerous papers and reports and a collection,* The Vegetation Continuum: A Collection of Readings *(with W. R. Keammerer; INSTAAR, 1973). He is currently a University of Colorado Faculty Fellow (1978-1979).*

Roger G. Barry, professor of geography with the Institute of Arctic and Alpine Research at the University of Colorado at Boulder, has conducted research on the climatology of arctic and alpine areas, climatic change, and synoptic climatology. He has done field work in the mountains of · Colorado and Papua, New Guinea, and in Arctic Canada and has published three books, most recently, Arctic and Alpine Environments *(coedited with Jack D. Ives; Methuen Ltd., 1974).*

W. Dwight Billings, James B. Duke Professor of Botany at Duke University, is a specialist in arctic, alpine, and desert ecology, physiological ecology and experimental biogeography. He is chairman of the Ecological Section of the Botanical Society of America, a member of the Scientific Advisory Committee of the Institute of Arctic and Alpine Research, and a member of the editorial boards of Arctic and Alpine Research *and* Ecological Studies. *He has published some 65 papers and books, including* Plants, Man and the Ecosystem *(Wadsworth, 1970) and* Vegetation and Environment *(Vol. 6, Handbook of Vegetation Science; edited with B. R. Strain; The Hague: Dr. W. Junk Publishers, 1974).*

Robert F. Grover, professor of medicine in the Division of Cardiology at the University of Colorado School of Medicine, is also director of the Cardiovascular Pulmonary Research Laboratory and of the High Altitude Research Laboratory His fields of expertise are cardiovascular and respiratory aspects of exercise and high altitude physiology. A fellow of the American College of Cardiology, the American College of Chest Physicians, and the American College of Sports Medicine, he is the author of some 140 books, papers, and reviews.

Jack D. Ives, professor and director of the Institute of Arctic and Alpine Research at the University of Colorado at Boulder, specializes in geomorphology and geoecology. He is chairman of the IGU Commission on Mountain Geoecology, and of the U.S. Directorate for a "Man and the Biosphere" project studying the impact of human activities on mountain ecosystems. A founder of the quarterly journal Arctic and Alpine Research, *he is chairman of its editorial board and is the author of some 60 papers in this field, as well as coeditor of* Arctic and Alpine Environments *(with Roger G. Barry; Methuen Ltd., 1974).*

Daniel H. Knepper, a geologist with the United States Geological Survey in Denver, was previously a research associate with the Institute for Arctic and Alpine Research and assistant professor of geology at the University of Colorado. His main areas of interest are remote sensing and structural geology.

Malcolm Mellor is a research civil engineer and research physical scientist at the Cold Regions Research and Engineering Laboratory in Hanover, New Hampshire, and an engineering research consultant. He is secretary of the International Commission on Snow and Ice, a member of the Committee on Glaciology (U.S. Polar Research Board), and a member of the editorial board of the Journal of Glaciology. *He is the author of over 100 publications, including many on the physics and mechanics of ice and snow.*

R. Brooke Thomas, associate professor of anthropology at the University of Massachusetts in Amherst, specializes in human ecology. He is a member of the U.S. Man and the Biosphere Directorate for assessing the impact of human activities on mountain ecosystems, and his publications include "Human Adaptation to a High Andean Energy Flow System" and "Geoecology of Southern Highland Peru: A Human Adaptive Perspective" (with B. Winterhalder).

High Altitude
Geoecology

Introduction and Commentary

Patrick J. Webber

Background

When I was first approached by the program committee of
the 1977 annual meeting of the American Association for the
Advancement of Science (AAAS) to organize a symposium, I was
asked to consider a theme dealing with the biology of high
mountains. Because the meeting was to be held in Denver,
Colorado, the theme seemed appropriate enough, but when I
learned a little later that all sessions of the meeting would
have as a central theme *Science and Change: Hopes and Dilemmas*,
I felt the symposium must at least deal with man in the high
mountains. With this in mind, and being cognizant of the
increasing involvement of my home base, the University of
Colorado's Institute of Arctic and Alpine Research (INSTAAR),
in a fledgling international program concerning man and mountains (1), the symposium reported in this volume began to
take shape.

The international program to which I refer is the Man
and the Biosphere (MAB) Programme which is being sponsored
and promoted by the United Nations Educational, Scientific,
and Cultural Organization (UNESCO). It is intended that
MAB should function as an intergovernmental, interdisciplinary, problem-solving effort of many years duration and will
focus on four areas (2):

(1)   the general study of the structure and functioning
      of the biosphere and its ecological regions;

(2)   systematic observation of changes brought about by
      man in the biosphere;

(3)   the study of the effects of these changes upon
      human populations; and

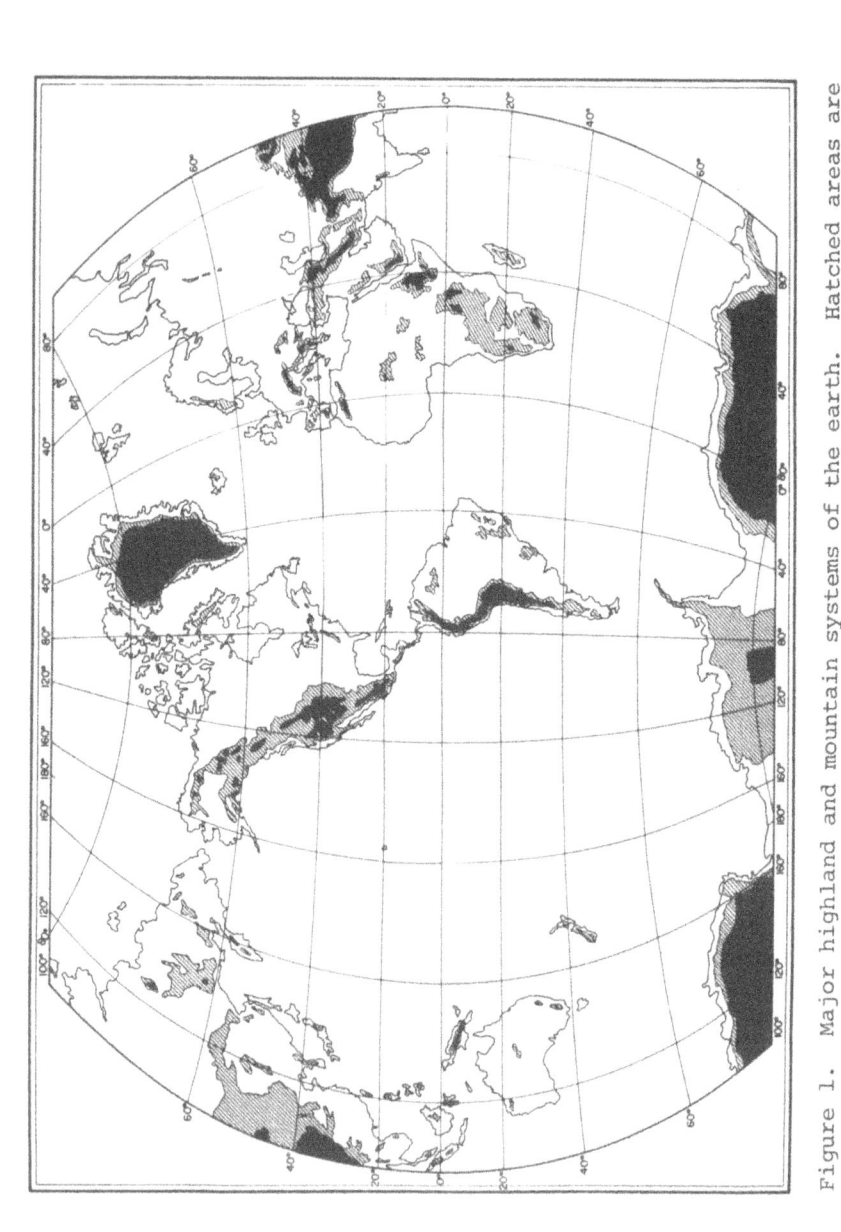

Figure 1. Major highland and mountain systems of the earth. Hatched areas are 1000 to 2000 m above sea level while solid areas are above 2000 m.

(4) the education and information needed on these sub-
     jects.

Fourteen project areas have been designated for study; one of
them is Project 6 (MAB-6): impact of human activities on
mountain and tundra ecosystems.

In the United States MAB-6 is a fledgling program (3).
It was hoped that this symposium would serve as a state of
knowledge indicator and as an expression of our perception of
man's role in mountain and tundra ecosystems; it might even
serve as a catalyst to aid the work of a MAB-6 effort in the
United States.

The symposium was called *High Altitude Geoecology*.
Geoecology is a term defined by Carl Troll (4) and can be
equated with landscape ecology; because humans are an inevitable part of the modern landscape, their role is implicit in
the definition. Thus the topic of high altitude geoecology
would be relevant to MAB-6, although the emphasis was intended
to be on mountains primarily, and only incidentally on tundra.
The book is presented as a contribution to the work of MAB-6
and the International Geographical Union Commission on Mountain
Geoecology. This commission was founded by Carl Troll (5).

Depending upon definitions, the extent of mountainous
regions of the earth will vary. From Figure 1, in which the
principal land surfaces exceeding 1000 m above sea level are
indicated, one can estimate that somewhere between one-third
and one-quarter of the earth's land surface may be classified
as mountainous. Actually 29% of land surfaces are above
1000 m (6) and of these 12% are alpine or tundra (7). Approximately 200 million people directly depend on these lands for
their livelihood (8). Although these people may represent
only about 6% of the world population (9), an even larger
number live in the surrounding lowlands and are affected by
the mountains. The effect of mountains on surrounding lowlands is primarily through climatic modification. The presence of mountains controls wind, temperature, and precipitation patterns. Mountains supply many important resources
such as pastures, forests, minerals, water, hydroelectric
energy, and even recreational opportunity to those who live
in them and visit them. One could continue waxing about a
potpourri of characteristics concerning mountains, for example their source of inspiration to poets and sages. They contain the few remaining remnants of unexplored land, and much
basic knowledge of the natural history of earth has been re-
vealed through the study of mountains. It is imperative in
a finite world that man should strive to understand and stew-
ard this vital segment of the planet earth.

The Symposium

The temporal constraints of the symposium allowed for seven, forty-minute contributions and a discussion session. Seven contributors were approached and asked to prepare a statement on their field of expertise, bearing in mind the MAB rationale. They were asked to address themselves to mountains in general rather than to a specific mountain region. The sequence of papers in this volume is, with one exception, that of the symposium--the contributions of Knepper, (remote sensing - Chapter 3) and Barry (climate - Chapter 4) have been reversed. As with the majority of symposia, all facets of a theme cannot be addressed and the treatment of an assigned topic must be left to the discretion of the contributor. It is my opinion that the written versions of the contributions, as presented here, faithfully reflect the original oral presentations except for the loss of much visual materials. The symposium itself, on the basis of subsequent audience response and involvement in the final discussion, was judged to be successful. It is to be hoped that this written version will be useful to people who were unable to attend.

The lead paper was presented by J.D. Ives (Chapter 2) who perhaps has done more than anyone in the United States to promote simultaneously the related causes of MAB-6 and Troll's concept of geoecology. The paper details the problems confronting man in the high mountains and considers what the scientist must do to preserve the mountains. Ives takes three contrasting case histories to illustrate his thesis-- that an integrated, interdisciplinary, and international study of past and present mountain processes is needed and that only this along with education and concerted action can preserve mountain ecosystems. The first case history concerns the problem of overpopulation and migration from the tropical high Andes and the Himalaya, the second the problem of tourist impact in the Austrian Alps, and the third the legislative problem relating to construction in areas of natural hazard in the rapidly developing central Rocky Mountains.

The need for detailed knowledge of the mountain environment leads naturally to the next three papers. D.H. Knepper, a geologist with a strong grounding in the principles of remote sensing, emphasizes the need for innovative techniques to make the necessary inventories of these often inaccessible areas. He discusses the types of data that can be obtained with modern methods and stresses the need for the data gathered to understand the theory of remote sensing methods and the limitations of the accumulated data. The reader who is interested in pursuing the topic of Chapter 3 would do well to

read sections of an earlier reference concerning a U.S. MAB-6
workshop (3). Here R.N. Colwell was the force behind a section
on applications of remote sensing to the management of high
mountains and tundra (3, pp. 13-15). Another paper by Colwell
is also very useful (10). In Chapter 4, R.G. Barry points
out that the climate of high mountains is a rather neglected
field. Barry is a climatologist whose qualifications are
attested to by his well-known textbooks (11, 12). He examines
the major elements of climate and discusses how these vary with
season, altitude, latitude, and topography and identifies
critical gaps in the understanding of mountain climates.
These relate to solar radiation, wind, and moisture balance.
Barry advocates an expanded program of climate measurement
in mountain environments to aid in the solution of air quality and energy problems as well as basic scientific questions.
The third environmental paper (Chapter 4) is by M. Mellor.
It was to many members of the audience the most remarkable
paper of the symposium, because of its detail about one topic:
water in its solid state in the high mountain environment.
Mellor's background as a civil engineer with considerable
experience, possibly second to none, in the applied science
of snow and ice has provided a definitive state of knowledge.
This concept of organizing knowledge prior to new research
programs is critical. Too rarely is this done and we should,
as MAB-6 develops, do more of this.

Initially I had hoped to snare both a plant and an animal ecologist into contributing to the symposium. However,
because of time constraints, I decided one or the other would
have to do. Hindsight now permits me to realize I need not
have worried, because W.D. Billings, the plant ecologist
I invited, was able to deal with the deficiency in superb
fashion by ably incorporating nonphotosynthetic organisms
into his discussion. In Chapter 6 Billings shows his mastery
and comprehension of alpine ecosystems. In fact, it is a
fine demonstration of practicing his own teaching concerning
the holocoenotic nature of ecosystems (13, 14). By discussing the evolution and structure of alpine ecosystems, he
singles out the characteristics which high mountain systems
have in common. This list of physical, biological, aesthetic,
economic and scientific characteristics serves to focus an
awareness of what must be done to protect the mountain ecosystem. In fact, the research priorities summarized in this
chapter should form the basis of any MAB-6 program.

The remaining papers deal with man. Chapter 7 by Grover
considers the physiological effects of living and working at
high altitude. Grover is both a physician and scientist and
is well known for his work concerning biomedical problems
induced by the reduced partial pressure of oxygen (15).

He writes with easy style and explains why, even with acclimation, man cannot entirely overcome the effects of high
altitude and if he wishes to be there then there is a minor
price to pay. Further information on human physiology of
high altitude is available in a recent book, *Man in the Andes*
(16), which was the result of one of the International Biological
Programme (IBP) projects. IBP was the forerunner of MAB and
Man in the Andes (16) provides a fine heritage for MAB-6.
The final paper of the symposium was given by R.B. Thomas who
participated in the IBP Andean study (17, 18). It is fitting
that Chapter 8 is the longest chapter in the book because
it raises the most serious problem of high altitude geoecology
and of the MAB-6 effort. Thomas has combined an understanding
of ecological principles with an understanding of human nature
to show that native peoples will suffer severe cultural and
physical harm as they experience the effects of competition
from lower elevations. Thomas ended his paper and chapter with
a poem so poignant that it surely should move us to better
understand the high altitude peoples and their environment.

At this point the symposium participants were finally
faced with the theme of the 1977 AAAS meetings; *Science and
Change: Hopes and Dilemmas*. Certainly we should protect
mountain people and their environment, but the development
of both is unavoidable. The challenge will be to both protect and develop in a harmonious fashion.

Acknowledgments

I should like to thank the contributors to this volume
for participating in the symposium and for preparing their
manuscripts so thoroughly. Dr. Harold Steinhoff, Professor
of Wildlife Biology, Colorado State University, kindly and
expertly chaired the second half of the symposium and provided many useful observations based on his experience.
Support and facilities for the preparation of the volume
were provided by INSTAAR at the University of Colorado.
Ms. Kathleen Salzberg provided considerable editorial assistance and quality control during the preparation of the manuscripts. My greatest thanks, however, go to Ms. Laura Kohn
who typed the camera-ready copy from which the book was prepared. Finally I wish to express my thanks to the staff of
the AAAS Publications Office who treated their tardiest symposium editor with such extraordinary patience.

References

1. J.D. Ives, The Unesco Man and the Biosphere Programme
    and INSTAAR. *Arct. Alp. Res.*, 6(3): 241-244 (1974).

2. Unesco, Programme on Man and the Biosphere (MAB), International Co-ordinating Council of the Programme on Man and the Biosphere (MAB), First session, Paris, 9-19, November, 1971, Final Report, *MAB Report 1*, Unesco, Paris: 65 pp. (1972).

3. J.D. Ives and A. Stites (eds.), Unesco Programme on Man and the Biosphere (MAB) Project 6: Proceedings of the Boulder Workshop, July 1974, Colorado. *INSTAAR Spec. Publ.*: 122 pp. (1975).

4. C. Troll, Geoecology and the world-wide differentiation of high-mountain ecosystems. *In* Troll, C. (ed.), *Geoecology of the High-mountain Regions of Eurasia.* Erdwissenschaftliche Forschung 4, Steiner, Wiesbaden: 1-16 (1972).

5. J.D. Ives, Preface. *Arct. Alp. Res.*, 10(2): 159-161 (1978).

6. A.N. Strahler, *Introduction to Physical Geography.* Wiley and Sons, New York: 455 pp. (1954).

7. L.E. Rodin and N.I. Bazilevich, *Production and Mineral Cycling in Terrestrial Vegetation.* (English translation by G.E. Fogg.) Oliver and Boyd, Edinburgh: 288 pp. (1968).

8. J.D. Ives, The development of mountain environments: Munich international workshop. *Arct. Alp. Res.*, 7(1): 101-102 (1975).

9. R.F. Dasmann, *Environmental Conservation.* Wiley and Sons, New York: 436 pp. (1976).

10. R.N. Colwell, Some significant elements in the new remote sensing panorama. *Surveying and Mapping*, 34(2): 133-142 (1974).

11. R.G. Barry and R.J. Chorley, *Atmosphere, Weather and Climate.* Methuen, London: 379 pp. (1971).

12. R.G. Barry and A.H. Perry, *Synoptic Climatology: Methods and Applications.* Methuen, London: 555 pp. (1973).

13. W.D. Billings, *Plants, Man, and the Ecosystem.* Wadsworth, Belmont, California: 60 pp. (1964).

14. W.D. Billings, Environment: concept and reality. *In*
     B.T. Strain and W.D. Billings (eds.), *Handbook of
     Vegetation Science Part IV: Vegetation and Environment.*
     Junk, The Hague, Netherlands: 9-35 (1974).

15. R.F. Grover, Man living at high altitudes. *In* J.D. Ives
     and R.G. Barry (eds.), *Arctic and Alpine Environments.*
     Methuen, London: 817-830 (1974).

16. P.T. Baker and M.A. Little, *Man in the Andes: A Multi-
     disciplinary Study of the High Altitude Quechua.* Dowden,
     Hutchinson, and Ross, Stroudsburg, Pa.: 482 pp. (1976).

17. R.B. Thomas and B.P. Winterhalder, Physical and biotic
     environment of southern highland Peru. *In* P.T. Baker
     and M.A. Little (eds.), *Man in the Andes: A Multidisci-
     plinary Study of High Altitude Quechua.* Dowden,
     Hutchinson, and Ross, Stroudsburg, Pa.: pp. 21-59 (1976).

18. R.B. Thomas, Energy flow at high altitude. *In* P.T. Baker
     and M.A. Little (eds.), *Man in the Andes: A Multidisci-
     plinary Study of High Altitude Quechua.* Dowden,
     Hutchinson, and Ross, Stroudsburg, Pa.: pp. 379-404 (1976).

Applied High Altitude Geoecology

2

Can the Scientist Assist in the Preservation of the Mountains?

Jack D. Ives

Introduction

High altitude geoecology, as an important subdiscipline within the natural sciences, probably owes more to the inspiration and lifetime research of the late Professor Carl Troll (1900-1975) than to the work of any other single person. One of the crowning organizational achievements of Troll's long career was the foundation, in 1968, of the International Geographical Union (IGU) Commission on High Altitude Geoecology. Troll carried forward the classic German approach to mountain geography by considering high mountains in a four-dimensional intellectual setting and by developing research methods to reflect this--the dynamism of mountain landscapes in space and time. An important corollory was his demonstration for the need for an interdisciplinary approach. We might still ask, however, what is *geoecology*, in general, and *high altitude geoecology*, in particular? In a somewhat simplified sense, and perhaps etymologically, it can be explained as *ecology*, with a reemphasis on the *geosciences* of geography and geology. The word *ecology* alone should be adequate except that, at least in the English-speaking world, ecology is a biologist's discipline, and in popular usage, has become synonymous with *environment* (incorrectly, of course). Thus I like Troll's word, *geoecology*, its use emphasizing the need to treat biotic and abiotic aspects as equals in an interdependent relationship, and its application to the high mountains of the world--defined as land with sufficient relief that multiple life zones, or altitudinal belts, are encompassed--is no more than a reflection of his own particular vocation. But more than this is required. Man, as an integral part of geoecology, is frequently the victim of neglect on the part of the geo-ecologist, or ecologist, preferring to study so-called "natural" systems. The Man-Mountain relationship constantly needs emphasis. I am, therefore, proposing to extend Troll's usage of high altitude geoecology to bracket the natural and human

sciences within the context of the high mountain systems of
the world.   I have used the term *applied* in the title of this
paper because of my personal conviction, shared by many
colleagues, that mountain specialists, and generalists, have
an obligation today more than ever before to apply their
knowledge, their intuition, and their moral strength to the
task of achieving a better balance between late-20th century
society and high mountain landscapes, before we witness the
destruction, within our lifetimes, of a large and vital part
of our heritage and of the natural resources upon which we
will increasingly depend for survival. By this I am not
urging that all mountain scientists devote all their energies
to technological research programs.   There are massive gaps
in knowledge in all of the classical disciplines that need to
be filled before fully intelligent planning of mountain
landscapes can be achieved.   What I am requesting is a widespread willingness to contribute one's knowledge and experience
to a process whereby planner, politician, *habitant*, and
scientist work together to achieve a better man-environmental
balance.

From an organizational point of view the formal development
of high altitude geoecology is remarkably recent, although it
has been practiced informally since early in the present
century--Russian workers, for instance, for decades have made
studies in biocenology.   But, as mentioned already, under the
influence of Carl Troll's international prestige, the Commission
on High Altitude Geoecology was founded by the General Assembly
of the IGU at its meeting in New Delhi in November 1968.   The
group of scientists that Troll then brought together was both
international and interdisciplinary in scope, although with a
concentration on the natural sciences.   Special emphasis was
originally placed upon studies of the world-wide phenomena of
the upper timberline. Necessitated by a serious illness,
Professor Troll relinquished the office of commission chairman
in 1972 and did me the honor of nomination to succeed him for
the commission's second four-year term following the Montreal
Congress of the IGU, although he remained a very active
member of its executive committee until his death.   This
heavy involvement on my part led to my invitation to an
organizational meeting of the Unesco, Man and the Biosphere
(MAB) Program, Project 6:   study of the impact of human
activities on mountain and tundra ecosystems.   This meeting,
at the invitation of the Austrian Government, was held in
Salzburg in January/February 1973, where initial plans were
prepared for development of an applied research program for
the high mountains of the world.   This was followed in November
1973, by a further planning meeting in Lillehammer, Norway.
On this occasion Unesco established an International Working
Group for MAB Project 6 which in turn initiated successive

regional meetings in Vienna, Austria (December 1973); La Paz, Bolivia (June 1974); Boulder, Colorado (July 1974); Kathmandu, Nepal (October 1975); Bogota, Columbia (June 1976); and Briançon, France (May 1977). These meetings have developed international and regional strategies and have encouraged a large amount of research activity, although the full impact of the planning still awaits the allotment of significant levels of national and international funding.

From the point of view of the formal organization of mountain geoecology, the IGU General Assembly, during its Moscow Congress of July 1976, agreed to the transformation of the original Commission on High Altitude Geoecology, now that its second statuatory term had expired, into a new Commission on Mountain Geoecology. This transformation was requested so that the new Commission could take on a more fully applied role, and could relate formally to the Unesco MAB Project 6. The following text is a discussion of what I consider to be some of the main challenges facing mountain geoecology, and an outline of some of the most important destructive processes occurring in our high mountains today. As with all complex processes involving Man and his natural resources, the issues are by no means restricted to the scientific sphere, although our basic knowledge is still dramatically inadequate. The most difficult obstacles facing any real degree of success in developing comprehensive and ecologically and socially sound planning policies are probably political. Nations must work together and concentrate more of the defense of their natural and cultural resources rather than on so-called "defense" in the traditional sense of more and better offensive weapons; resource developers must see more in their long-term objectives than "profit," whether in the capitalistic or communistic definition; and we must find a way to work with the mountain peoples so that their traditions and extensive experience of environmentally sound strategies are incorporated into the new methods of land use and land management to achieve something approaching sustained yield, determined on a very long-term basis.

The main part of this paper, which was originally written as the introductory chapter for a book on the environmental problems of the Himalaya (1), now follows. Its writing has been largely based upon my experience with the development of the Unesco MAB Program, Project 6, and with the work of the IGU Commission on Mountain Geoecology and its forerunner. It is an attempt to synthesize the contributions of many collaborators, but, of necessity, only provides coverage of a few highlights. Those interested in the details of the planning meetings should consult the numerous publications of Unesco and the Commission. The formal objectives of the Commission

for 1976-1980 are to be found in publications listed in the
References for this paper (2,3,4,5,6,7).

The Problems of the High Mountains

The problems of increased population pressures and
environmental degradation facing the world's great mountain
regions are no more, and indeed no less, than those facing
the entire planet. But in the mountains there is one great
difference, if only a difference of magnitude, which is
induced by the basic topographic characteristics of mountains--
high altitude and slope! In effect, soil degradation, infer-
tility, and abuse of forest and grazing lands by destructive
over-use will have drastic consequences anywhere. On mountain
slopes, however, they cause a rapid increase of slope instability, which is a polite way of saying soil erosion, gullying,
landslides, increased irregularity in the hydrological
cycle. If this was all, the problem would be severe enough,
but perhaps not of earth-shattering consequence: mountain
lands would suffer damage, perhaps irreparable damage, and
the people who live there, a tiny fraction of the total world
population, would suffer, and an elitist group of mountaineers,
trekkers, and tourists would be upset and would have to seek
their pleasures elsewhere. Consider further, however:
geomorphologists use the expression "high energy environment"
when discussing geomorphic processes--an academic expression
related to the terms "soil erosion" and so on, used above.
Now a high energy environment is defined as an area possessing
high relief, where processes--natural, man-made, or man-
augmented--are rapid. In other words, these processes are
influenced by a combination of elevation, slope angle, and
gravity. The geomorphic equation includes source → erosion →
transport → deposition. Usually the point of deposition is
the lowlands, plains, and valleys, and embayments of the
ocean, beyond the mountains, precisely those areas that
provide the life-support base for hundreds of millions of
people. Thus the tragedy of the mountains is the tragedy of
the plains, and of the entire world.

From this purely practical viewpoint, those people
devoted to preservation of some semblance of man-environment
balance in the mountain lands are facing a task of the
greatest imaginable magnitude. Politics, religion, husbandry,
technology, land ownership, social relations, economics, and
high finance as well as natural and human sciences research,
are all involved. The problem cannot be discussed in a
single paper; it can hardly be conceived of by a single
author. Here an attempt will be made to examine basic research
needs in the natural and human sciences, needs that must be
at least partially fulfilled before adequate long-range

planning, reconstruction, resettlement, reafforestation, and
preservation of much of that which is still viable, or partly
so, can be provided for. This paper is not intended to argue
the needs solely along the practical lines emphasized so far.
What of the aesthetics of mountains? What of the inspirational values, perhaps intangible, yet surely of inestimable
worth? So many of the greatest facets of the human spirit
are enhanced by the inspirational contribution of our mountains.

Gustav Mahler had the habit of retreating to his beloved
Alps, where he composed much of his great music. In an
infinitely more humble mode, this writing is taking shape on
a ridge crest looking onto the Bernese Oberland, coral pink
as the sun sets on the new snow of early winter. The Psalmists
immortalized the process of lifting up one's eyes to the
hills and the Puranas glorify the mountains:

> As the sun dries the morning dew,
> So are the sins of man dissipated
> At the sight of Himalaya

The preceeding eulogies might be passed off as the
reflections of yet another elitist minority. But this is
surely not so. I describe a personal experience--that of
sitting on the doorstep of a humble house in Darjeeling with
a Tibetan friend. It is close to sunset; heavy cloud, with
great turrets of cumulus pushing up from the valley floor,
had veiled Kangchenjunga throughout the day; now the cloud
tops had flattened and lowered, so that the mountain floated
on a sea of vapor as the sun sank into it, turning everything
blood-red, gold, yellow, purple, and stark green. Tshering
Choni turned to me and explained that, despite his poverty,
he recognized that he was one of a privileged few amongst the
world's total population who could have their spirits uplifted
by such an experience. It is surely the worst kind of arrogance
for us to assume that only those fortunate enough to have had
a special education are equipped to appreciate the beauty of
the Himalaya, a beauty that can only be preserved in the
fullest sense by the achievement of a functioning balance
between man and mountains.

The problem of the Himalaya, and of many of the world's
mountain ranges, can be stated very simply. In an already
high energy environment, rapidly accelerating human pressures
over the last 50 years are exponentially augmenting massive
and rapid transfer of large volumes of material downslope
overcoming, at least in part, Nature's counteracting forces
of slope afforestation. The counteracting forces of Nature,
the natural plant associations, and the hitherto reasonably
well balanced agricultural associations are being disrupted;

soil fertility is being reduced; and whole hillsides are
being lost to productive human endeavor. Thus we have a
human element imposed upon a natural system. We cannot begin
to quantify the human element and thereafter move on to
develop long-range planning and other counter measures without
understanding more clearly the natural system, or the man-
nature interacting system that existed prior to the middle of
this century. Nor can we call a halt to the devastating
human process, even if governments and planners possessed
absolute dictatorial powers. Nor can we delay the necessary
decision making until all the research needed has been
completed.

Examples of environment-management problems from three
different mountain regions will now be outlined. They are
deliberately selected from strongly contrasting areas, with
the contrast arising from differences in culture, length of
human occupation, and physical geography. Thus I will
discuss out-migration from the Altiplano of the Tropical
Andes, advanced two-season tourism in the Austrian Alps, and
construction of second homes and resorts in the path of
natural hazards in the Colorado Rocky Mountains. These are
intended to illustrate that mountain landscape deterioration
is not unique to tropical and subtropical high mountains, nor
is it concentrated in heavily populated areas, nor in those
that have a long history of human occupation. It is a world-
wide phenomenon, equally rampant in rich and poor countries
alike, that has only been recognized during the past decade.
Indeed, the magnitude of the problem has been hardly identified.
Thus it may be of some value for Himalayan planners to understand how widespread and complex the phenomenon is, and how
vital it is for efforts to be made to share growing experience
from different areas. The awareness of a slowly growing
international mountain effort will hopefully prove of both
practical and moral assistance for those shouldering the task
of saving and repairing the world's greatest mountain system,
both for the peoples living there, and for the growing number
of visitors, and especially for the hundreds of millions
living downslope whose fields and homes are being inundated
with silt and water.

The Tropical High Andes:
Overpopulation and Out-Migration

The Andes stretch over 70 degrees of latitude along
the western margin of South America making up one of the most
uniquely varied ecological structures of the biosphere. Two
themes are central to this mountain system: its youth and
massiveness, both vertically and geographically. Immature
soils and recent biogeographical development are given added

complexity by rugged topography and close proximity of diverse ecological zones that range from tropical rain forest to desert, and from coastal to alpine. The individual mountain ranges and intermontane valleys comprise a patchwork of habitats that combine to form an environment that is heterogeneous in time and space. This diversity and three dimensional extent provides many comparisons with the Himalayan system.

Any summary of the highland climate of the Tropical Andes, for instance, must emphasize scale and environmental pattern. For each of the major climatic elements, attention must be paid to the general pattern as well as the sources of variability that introduce regular and nonregular alterations. Such sources include altitude and topography, and meteorological changes which occur daily, monthly, and annually. When these elements are superimposed and interacting, a finely scaled mosaic of environmental conditions emerges. One valley has different weather from the next, the valley floor contrasts with the surrounding slopes; on a single slope, the banks of a small stream or shelter of a rock outcrop, each possess unique microclimates or microecosystems. The availability of moisture and protection from frost make these minute changes critical to plants and animals, and to the human groups depending upon them. Thus caution must be taken when classifying Andean environments into the broad altitudinal belts of selva, montana, and sierra, there is the risk of overlooking significant environmental constraints and effective responses to them.

One way of organizing our perception of the tropical high Andean environment is to consider its influence on plants and animals. Specifically, we can delineate those conditions in the highlands to which any population of the biotic community must make some kind of adjustment in order to function and reproduce effectively. In biological terms we speak of environmental constraints which are identified as either stressors, or limiting factors, and of adaptions, or the variety of genetic and nongenetic responses that plants and animals use to mitigate the constraints. Potential environmental constraints of the Altiplano can be grouped into five categories (see Thomas, this Volume, ref. 29):

(1)   reduced partial pressure of oxygen and carbon dioxide as a function of altitude, low absolute vapor pressure; high background radiation;

(2)   rugged topography and poorly developed soils with marginal availability of certain nutrients;

(3) low temperatures with pronounced diurnal variation,
     and frequent and intense frosts which can occur
     in any season;

(4) a lengthy dry season and irregular distribution of
     precipitation; droughts which may last several
     years and are unpredictable;

(5) a biotic community with limited productivity spread
     over wide regions.

Different properties of a stress will influence the
degree and kind of adaptation made. These properties can be
categorized as frequency, intensity, duration, and regularity.
The first three are self-explanatory; regularity means that
the stress varies according to a predictable and repeated
pattern. Thus lowered partial pressure of oxygen and carbon
dioxide, and low barometric pressure, are examples of stresses
which are constant (regular and of infinite duration) and
which vary in intensity uniformly with altitude. Topography
and soils change over centuries and have their predominant
variability in space; their intensity, frequency, duration
(extent), and regularity can be judged from soil maps.
Climatic stressors can be characterized by the seasonal or
daily rhythms of moderate intensity, and by the sporadic and
intense instances of stresses which vary in duration and
which are generally irregular. The interruption of a regular
diurnal temperature variation by an airmass frost (black
frost or *helada negra*) is an example of an infrequent,
nonregular, and intense stress that can last for several
days.

All of the above physical stressors exert an influence
on the spatial and temporal productivity of the biotic
communities, and on the flow of energy and materials between
populations. While the lengthy dry season, soil conditions,
and diurnal temperature variations impose fairly constant
limits on biota, it is the irregular stressors, such as the
dynamic frosts and droughts, which are primarily responsible
for fluctuations in productivity from year to year. Organisms
face a variety of stressors and stress characteristics, and
their response is a complex interplay between various kinds
of adaptations. Plants and animals must be capable of
surviving the infrequent and rigorous environmental conditions
as well as the more common, regular ones. These then are the
broad environmental constraints of the Altiplano to which
resident human groups dependent upon herding and farming must
adjust. Any environmental planning that does not take them
into consideration will be courting disaster. Also, while
environmental constraints of this nature will vary in intensity

and relative importance between different mountain regions, the same general understanding is required.

Human Utilization of the Altiplano

Unlike most other animal populations, human groups have the capacity to extensively modify their biotic environment. This has permitted some groups to achieve a relative balance with the ecosystems upon which they rely, while for others it has allowed levels of exploitation which ultimately lead to the impairment of the ecosystem functioning, thus reducing future capacity for human utilization. Today changes are occurring that are increasingly destroying many of these balances as a new set of values, and often external pressures, overtake the traditional forms of occupation, or as improved health care upsets the population balance. Yet we cannot categorically condemn all modification that leads to environmental degradation, especially since human groups are becoming increasingly dependent upon this mode of utilization. What is possible, however, is identification of ecosystems which are more or less resilient to degradation in the sense that they are able to recover rapidly. Similarly, it is important to determine how different forms of degradation influence ecosystem function, and which of these lead to very long-lasting or irreversible damage. Examples include excessive erosion from overgrazing, air- and waterborne toxic waste products, strip mining, and deforestation.

In addition, the consequences of such degradation on human groups must be thoroughly examined and land-use alternatives for these groups must be recognized. In areas where abundant land is available permitting the group to "move on," and where recovery is rapid, the consequences of degradation are less serious. Unfortunately, this option is rapidly being exhausted. A further consideration is the extent to which degradation can be reversed or its consequences mitigated through support from national governments. This is frequently a costly option which will interfere with other national priorities, or be completely beyond the financial capacity of a developing country. Thus, if the biosocial consequences of degradation are severe, and if alternative ways of utilizing the environment are unavailable, outside support would be necessary if a group was to remain intact. Emigration from the ecosystem would be expected under such circumstances. The capacity for a nation to extend long-term support and absorb migrants, therefore, becomes an important consideration in initiating changes where the environmental consequences remain uncertain.

In the tropical Andes, we are confronted with a diversity of human utilization patterns that parallels the natural patterns in complexity and interrelationships. Human settlements span the range from small isolated rural communities to large urban centers. Cultural affinities likewise range from the traditional values of the Andean native to multiple aspects of the national culture. The outcome of this mix has produced considerable variation in attitudes and in patterns of environmental utilization, affording an excellent opportunity to assess their relative merits. Despite such diversity, general patterns of resource utilization do emerge. Environmental fluctuation and perturbation in the high Andes are marked compared with many other regions. As a consequence, the environment is less predictable, necessitating a wider range of adaptive strategies for those groups dependent upon it. This suggests that programs which advocate the production of a single crop or domestic animal will be met with continued resistance, unless they can demonstrate that risks involved are considerably below existing ones, or that the profits compensate for frequent losses.

Possibly the most predominant resource utilization pattern in the Andes is related to the vertical arrangement of ecozones. This creates a situation whereby a wide variety of resources are available within a relatively short vertical distance. Since any one zone usually does not produce all essential resources, some form of exchange is necessary. In the highest ecozones, for instance, where herding is the principal economic activity, it appears that the production of food energy is insufficient. This condition is mitigated through the exchange of animal products to lower zones and to the national market in return for high calorie foods. Should such an exchange be terminated, or the price of the animal products become significantly devalued, these herding groups could probably not persist. Preconquest patterns in the Andes indicate that access to essential resources was obtained by group land rights in more than one zone. Such a system has been largely replaced by exchange between groups residing at different altitudes, and by markets. While construction of roads facilitates the flow of resources between areas, it also isolates communities formerly served by foot or animal trade and hastens their collapse.

The question of community collapse or abandonment poses a potentially serious problem with regard to environmental utilization. A substantial portion of the highland population is composed of indigenous or peasant groups which are to a varying degree dependent upon their immediate environment for subsistence. As such, they are marginally connected to their respective national economies and cultures, although this is rapidly changing. Many groups are altering the manner in which they interact with their environments through the adoption of new technology. Others are giving up their traditional way of life, not because it is less adaptive, but because it is perceived by themselves or others as less "progressive." Thus, exposure to an alternative social system is sufficient to cause the abandonment of adaptive strategies which have operated in approximate balance with the productive potential of the environment, and have supported a substantial number of human groups apart from the national economy and independent of government services.

Such abandonment is viewed with concern for several reasons. The most pressing consideration depends on the ability of other human settlements or urban areas to effectively absorb and support migrants, and to continue to do so for the duration of the migration period. A second and more long-term concern, focuses on the consequences of abandonment which result in a reduced capacity to utilize large areas of the ecosystem for human support. Knowledge regarding the environment has been accumulated by these population for centuries and it will not be easily reacquired or duplicated through a short period of scientific research. Information concerning adaptive strategies in these ecosystems is especially important since these are areas where the introduction of modern production techniques have not been, in many cases, overly successful.

In tropical mountain areas above 2500 m, for instance, altitude-related factors influence the species of plants and animals which man can utilize. Traditional lowland crops and domestic animals are of decreasing utility above this altitude so that at elevations above 3500 m, lowland forms are almost entirely replaced by high altitude species. Likewise, as sloping terrain increases and flat areas suitable for agriculture become less frequent, mechanized farming practices become less feasible.

It is, therefore, important to realize that traditional groups are not coming to more "developed" settlements with empty cups. These are people who have considerable understanding of the environment and how it can be effectively used to support human groups. Thus, information on adaptive strategies employed in the past and present can provide valuable input regarding alternative land-use practices. While we can anticipate that traditional groups will become less frequent in some areas, many will remain intact in somewhat altered form. Alternatives which they adopt will depend upon how they view the environment, their values and goals, and the efficacy of new alternatives in meeting these. By being able to assess a group's adjustment to its environment, we are in a position to identify less successful strategies and possibly recommend acceptable alternatives in overcoming environmental constraints. The value of this would be to maintain viable traditional groups under conditions where further abandonment could not be accommodated.

In contrast to more traditional groups which attempt to produce most of their essential resources, increasing pressure on these Andean ecosystems is being exerted by groups utilizing specific resources who are directly connected with national or international distribution networks. Mining, oil and gas exploration, power and storage dams, tourism, and recreation provide examples of these resources. Because material support for these groups is largely from outside the ecosystem, and their purpose in utilizing it is rather singular, their impact on the environment can be serious. Many of these inputs are magnified by ecological problems which arise due to the need to use a labor force in isolated areas. If the labor force is drawn from the traditional groups mentioned above, this in turn impairs the ability of these groups to function in a more traditional way, hastening their breakup. If the labor force comes from the "outside," the lack of long-term commitment to the area will lessen the concern for human or environmental damage since it can simply be left behind when the job is completed. In addition, the transportation networks set up to service these installations will become routes of migration for more traditional groups. While such alteration is inevitable, information is sorely needed as to how change can be conducted more effectively. Migratory routes, for instance, allow for prediction of areas where rapid population growth can be anticipated. This suggests that such roads should be located so that they pass through areas that can withstand and support heavy population pressure.

One of the overall objectives of man-environment research in the Andes is the optimization of human welfare while maintaining the integrity and balance of ecosystems and minimizing irreversible environmental perturbation. Human welfare in this context is to be understood as referring to the fulfillment of biological requirements, as well as psychological and social needs. Optimization of human welfare, of course, is a complex and difficult goal since it is dependent upon having sufficient knowledge in a variety of areas, including land-use alternatives, indicators of disruption, values and goals of local groups, and national priorities

An example of this point is an environment which has a number of land-use alternatives. It is expected that some of these would be unacceptable in terms of the goals and values of local groups. Likewise, the best alternative for local groups may not be considered in the national interest. Thus, a balance must be struck between local and national interests that is at the same time environmentally sound. Unfortunately, our knowledge of the integration of these data is rudimentary. Thus, in attempting to achieve this general objective, it will be necessary to establish a baseline of data which can be applied to a wide range of problems, and which will allow comparison between various settlement and land-use patterns. This is viewed as a core of information applicable to the analysis of most human groups. The following, therefore, is a brief statement of what are considered to be realistic research objectives which can be applied to the tropical Andes, and which will analytically lead to the aforementioned goal.

Objective I: Assessment of the Impact of Human Activities on the Environment. This program entails the detailed description of ecosystem structure and function, especially those aspects which are affected by human activities. General categories of information required are outlined below:

(1) Inventory of the System
      Geology, soils, meteorology and climatology,
        biota, human population

(2) Structure of the System
      Patterns of horizontal and vertical distribution
        of organisms
      Levels of species diversity and physiological
        adaptations of dominants
      Population dynamics of dominant species
      Genecology and evolutionary rates of native and
        crop species
      Phenology
      Biological and social structure of human
        populations

(3) Function of the System
      Primary production (biomass structure and growth)
      Secondary production
      Decomposition processes
      Nutrient cycling
      Optimum use--agriculture or watershed
      Man's interaction with the environment

It must be emphasized that if environmental description and prediction is to be accurate, it must be derived from a complete, extensive, and long-term data base which will yield intensity, frequency, duration, and regularity of common and rare environmental events. Without this our ability to identify environmental deterioration, or constraints on human groups, will be limited.

Objective II: Evaluation of the Ability of Human Groups to Meet Their Biological, Physiological, and Social Needs. Evaluation of the biological, physiological, and social well-being is an extremely valuable measure since it provides an indication of a group's success or failure in interacting with its environment. Especially important in this respect is the identification of groups that are unable to adjust to some aspect of their environment. A slow and prolonged individual growth pattern, for instance, may be advantageous in an environment where nutritional intake is low and inconsistent from year to year. There is a list of demographic, biological, and socioeconomic indicators, comparable in detail to those listed under Objective I.

Objective III: Prediction of the Consequences of Change on Both the Environment and Human Groups. Andean human populations and the environment are undergoing unprecedented change which is both positive and negative in its effect. Attempts are being made to increase and improve resource use, and to extend health delivery systems and education to larger segments of the population. At the same time, human groups are experiencing increasing population pressure, large and extensive migration, socioeconomic and technological upheaval, changing political and economic expectations, and abandonment of traditional resource practices. Evidence from practically all the Andean republics suggests accelerating damage to basic life-support systems, both in the highlands and adjacent downslope areas. These factors, in turn, are causing an ever increasing concern for man's biosocial well-being.

It is, therefore, extremely urgent to understand the extent and long-term impact of these changes. Objectives I and II provide approaches for assessing man's impact on the environment and its biosocial consequences before, in the course of, and after a change has occurred. Additional information needed in order to determine the characteristics of a change is listed below:

(1)   Identification of different types of change capable of producing serious disruption;

(2)   Conditions causing or contributing to change;

(3)   Rate at which change occurs;

(4) Short- and long-term consequences of a change;

(5) Environmental and human conditions existing before the change;

(6) Early indicators of disruption;

(7) Rehabilitation actions which are feasible and acceptable to local groups as well as the national government.

Focusing upon local or regional human populations, two broad types of change can be identified. First is that initiated by, or the result of, actions by the group in order to meet their requirements or goals. Second is external change introduced from outside the group. This may be unplanned (immigration) or planned, either to assist the group (health services, education) or to utilize local resources (mining, hydroelectric dams, tourism). In the case of unplanned change with serious consequences, preimpact surveys should be undertaken to provide the basis for determination of future disturbance and identification of vulnerable areas. For planned change, guidelines could be established as to the far reaching consequences of such actions. These could assist planners in avoiding actions leading to either impairment of the environment or the human well-being.

Ultimately it is desirable to develop a model which could be used to predict the consequences of foreseen and unforeseen changes throughout the tropical Andes. Needed to achieve this stage of predictability are the following:

(1) Identification of alternative land and resource utilization patterns within the research area;

(2) Determination of group values and goals with regard to their acceptance of land- and resource-use alternatives;

(3) Evaluation of national priorities being imposed on or accepted by groups, and the long-term consequences of these.

Probably the most important change taking place in the Andes in terms of magnitude, area affected, and overall seriousness is the movement of human populations. Emigration from highland areas results in part from growing population pressures on a limited resource base, and from real or perceived expectation of a better quality of life in other areas. While emigration alleviates such pressures in some

FIGURE 1

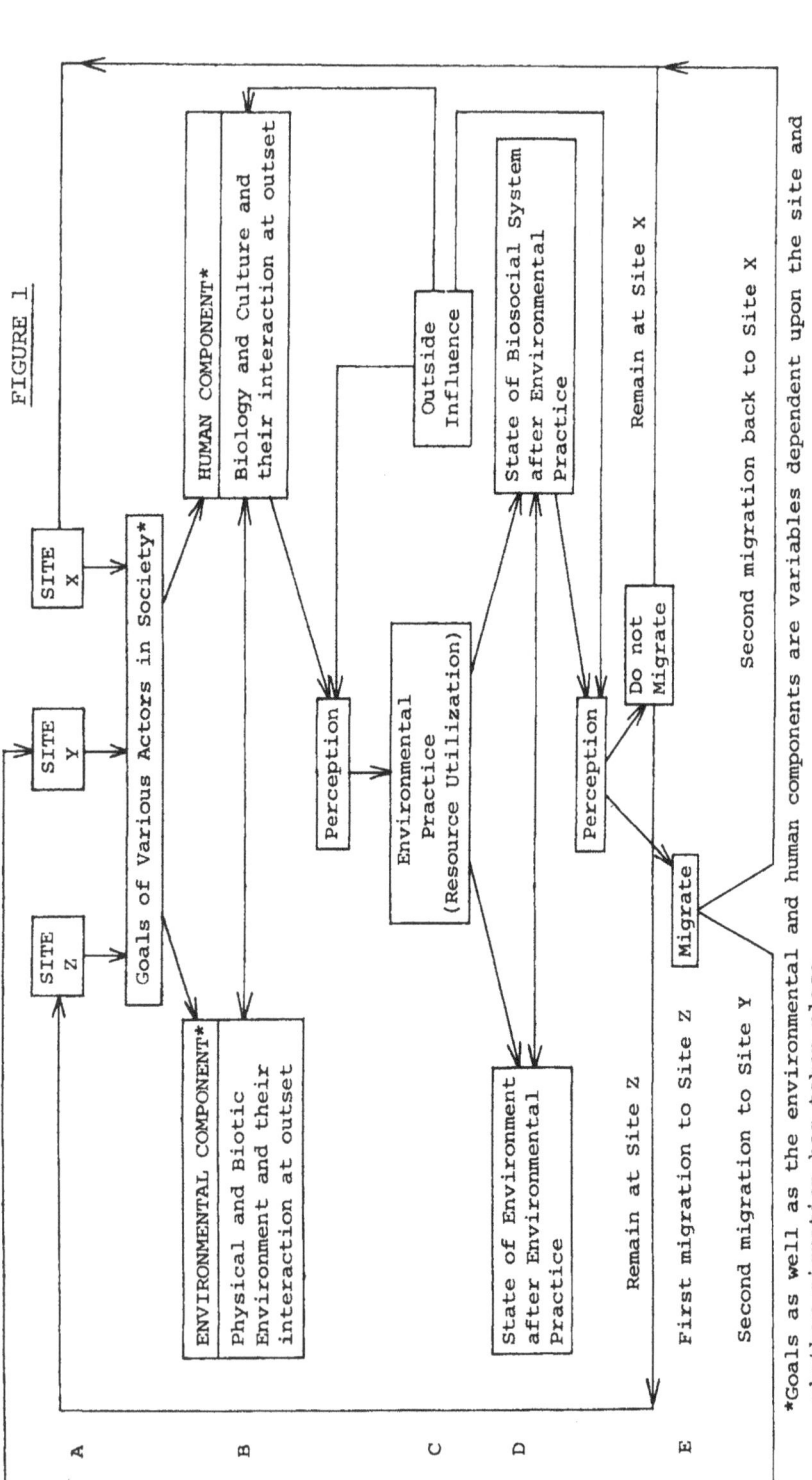

*Goals as well as the environmental and human components are variables dependent upon the site and whether migration has taken place

areas, in others it leads to underpopulation which, in turn, weakens exchange networks and can lead to community collapse. More serious is the pressure put on the life-support systems and the human populations in areas receiving the migrants. This is a problem of change which has countless secondary effects. It is, therefore, necessary to understand factors underlying the migration, its consequences, and the capacity of areas to absorb additional residents.

Conclusions

The foregoing description is but one small aspect of the developing applied research design for the entire Andean system. It was recognized that for the tropical Andes the most critical process occurring today is migration, both spontaneous and planned, which involves movement of large numbers of people along several pathways: to highland urban centers, to coastal urban centers, and to both coastal and eastern lowland rural areas. Part of the motivating force for this migration is population growth in the highlands which have limited environmental capacity for its absorption. The consequences are detrimental to both the receiving communities and the originating communities as well as to the environments in all areas. It was, therefore, decided that the most urgent research need was to concentrate on the migratory phenomenon. Figure 1 provides a schematic framework for investigating the impacts of this migration on the man-environment interactions. In addition, it was realized that because the research needs were vast, it was beyond the limits of likely available resources and man-power to tackle them on a broad scale. Thus, the approach was recommended of selecting a latitudinal transect across the Bolivian/ Peruvian Altiplano, from the Amazon lowlands to the desert of the Pacific coast. This would provide a basis for intensive inventory, human, and biotic energy flow studies and applications in the form of an action-oriented pilot project.

When successes from the pilot project become apparent, the approach could then more readily gather momentum and be applied to other critical areas. Similarly, the transect approach is recommended for the southern and northern Andes where problems other than migration are emphasized.

More information on the Andes is available in a multi-disciplinary study of the high altitude Quechua (8).

Figure 1 shows the general framework for investigating the effects of migration on man/environment interactions.

The Austrian Alps: Human Impacts in
a Two-Season Tourist Environment

This example is the product of the impact of tourism accelerating since about 1950 on a traditional farming community in the Eastern Alps. The environmental problem is complex despite the very small size of the settlement under consideration--Obergurgl, in the Austrian Tyrol. The objective was to devise an applied research model that would help to achieve a drastic change in land-use policy which appeared destined to destroy itself--to eliminate tourism through over development and environmental degradation. Because an early research design resulted in construction of an ingenious predictive model, now known as the "Obergurgl Model" this will be described in some detail.

In its broadest connotation, the Obergurgl Model can provide a vital unifying theme for many aspects of mountain research. It also symbolizes the progress that can be made through international collaboration and workshop experience. Its author, Dr. Walter Moser, introduced it initially as a very simple concept at the Unesco MAB-6 Salzburg panel of experts meeting (3), derived in part from a combination of International Biological Programme (IBP) ecosystem modelling experience and in part from a long personal involvement with the landscapes and people of Obergurgl itself, a type-example of tourist impact in the Austrian Tyrol. The initial concept was further developed at Lillehammer (4) and then systematized and put into working order at Schloss Laxenburg, May 1974, during an Alpine Areas Workshop held in conjunction with the International Institute for Applied Systems Analysis (IIASA). It remains to be seen whether such a model based upon an essentially small, simple Alpine tourist village with an unusual amount of available relevant data can successfully be developed and applied to larger, more complex tourist situations, or be used to help resolve other research problems, such as those involving technological impacts or combinations of competing land-use alternatives both in Europe and in other parts of the world. Nevertheless, the names Walter Moser and Obergurgl will remain an integral part of Unesco MAB Project 6.

The village of Obergurgl, at the head of the Ötztal at an altitude of nearly 2000 m in the Tyrolean Alps, faces problems similar to those of many of the world's mountain areas. Beginning about 1950, the village entered a period of economic growth driven by an apparently unlimited demand for tourism. This growth, expressed largely in terms of hotel construction, is beginning to have serious environmental consequences and further growth will soon be limited by

available building sites, if by nothing else. There is a key
simplification in the system: land ownership is tightly
controlled by a few families (originally farmers), and the
economic development rate is limited by the rate of human
population growth since this rate determines the number of
young people willing to invest in new hotels. Because
hotels can be built most readily on valley bottom land, land
which is most important for agriculture, an important land-
use alternative is being lost to the villagers. Obergurgl,
therefore, can be seen as a microcosm, albeit with some
elements missing and others exaggerated, of a major world-
wide problem: population and economic growth in relation to
diminishing resources--a problem especially critical in the
mountain world of steep slopes and limited level land that is
safe from natural hazards.

The focus of the Schloss Laxenburg Workshop was to
develop a preliminary model of human impact on a simple
alpine ecosystem and identify policy options through a
combination of the knowledge and insights of businessmen,
villagers, government officials and scientists. A prime
purpose became the use of the model as a device to identify
the potential areas of conflict and the missing critical
information, so that rational priorities could be set for
further descriptive and prescriptive analysis:

(1)   to promote communication among the various interest
       groups involved in Obergurgl studies, by using the
       simulation model to provide a common language and
       focus for attention;

(2)   to define, through data requirements for the model,
       critical research areas for MAB Project 6;

(3)   to provide tentative long-range (20 to 40 years)
       alternative forecasts for the people of Obergurgl
       concerning likely impacts of various development
       strategies that they consider practical.

A diagramatic outline of the first steps in the Obergurgl
Model is shown in Figure 2. The intention is to interrelate
the physical and human elements within the spatial framework.
For ease of exposition only a single dimension is used in
Figure 2; there are also regulating mechanisms which unite
elements in different frameworks.

Elements of the system, as shown in Figure 2, are also
greatly affected by external influences shown in Figure 3.
The elements are generalized and will need to be modified to
suit particular studies without affecting general comparability.

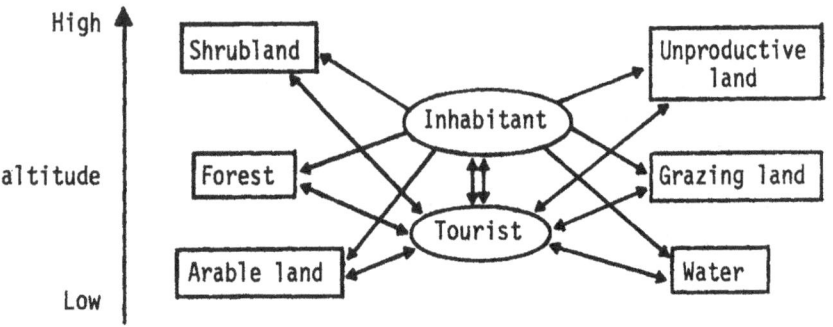

Figure 2.   A schematic framework for tourism research:
            internal sector.

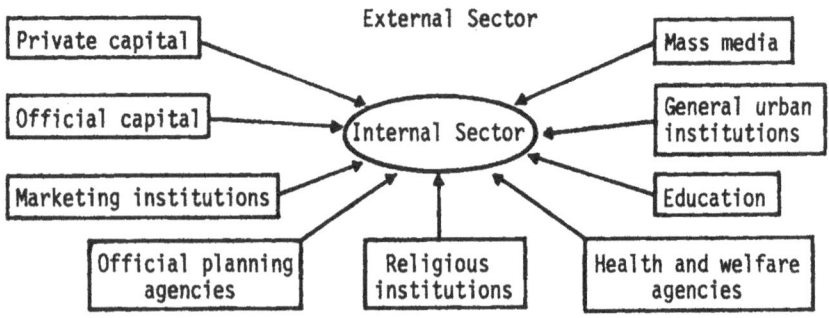

Figure 3.   Main elements of external sector impinging on
            internal sector.

The Schloss Laxenburg refinement of the model is presented as Figure 4. This provides a summary of the basic components and interactions in the model. The components were identified as the minimum set needed to make reasonable predictions for the next 30 to 40 years and fall into four major classes:

(1)   Recreational demand;

(2)   Population and economic development;

(3)   Farming and ecological change;

(4)   Land-use and development control.

Each class, or submodel, was then elaborated and further developed and finally used for production of thirty 50-year scenarios.

General Predictions

Though the model was developed to represent a rich variety of interactions and feedback mechanisms, its final predictions depend largely on a few key relationships. These can be summarized very simply:

(1)   In the face of essentially infinite potential demand, growth of the recreation industry has been limited by the rate of local population growth;

(2)   The amount of safe land for development is disappearing rapidly, while the local demand for building sites is continuing to grow;

(3)   As land is developed, prime agricultural land is lost and environmental quality decreases;

(4)   Recreational demand may begin to decrease if environmental quality deteriorates further.

Thus, Obergurgl may soon be caught in a painful trap, as its growing population and economy collide with declining resources and demand. This collision may be felt by the older, established hotel owners as well as the younger people if more hotels are forced to share a declining number of tourists.

An alternative future, again generated without development control, but assuming that recreational demand will remain at 1973/74 levels (e.g., continued energy and monetary crisis

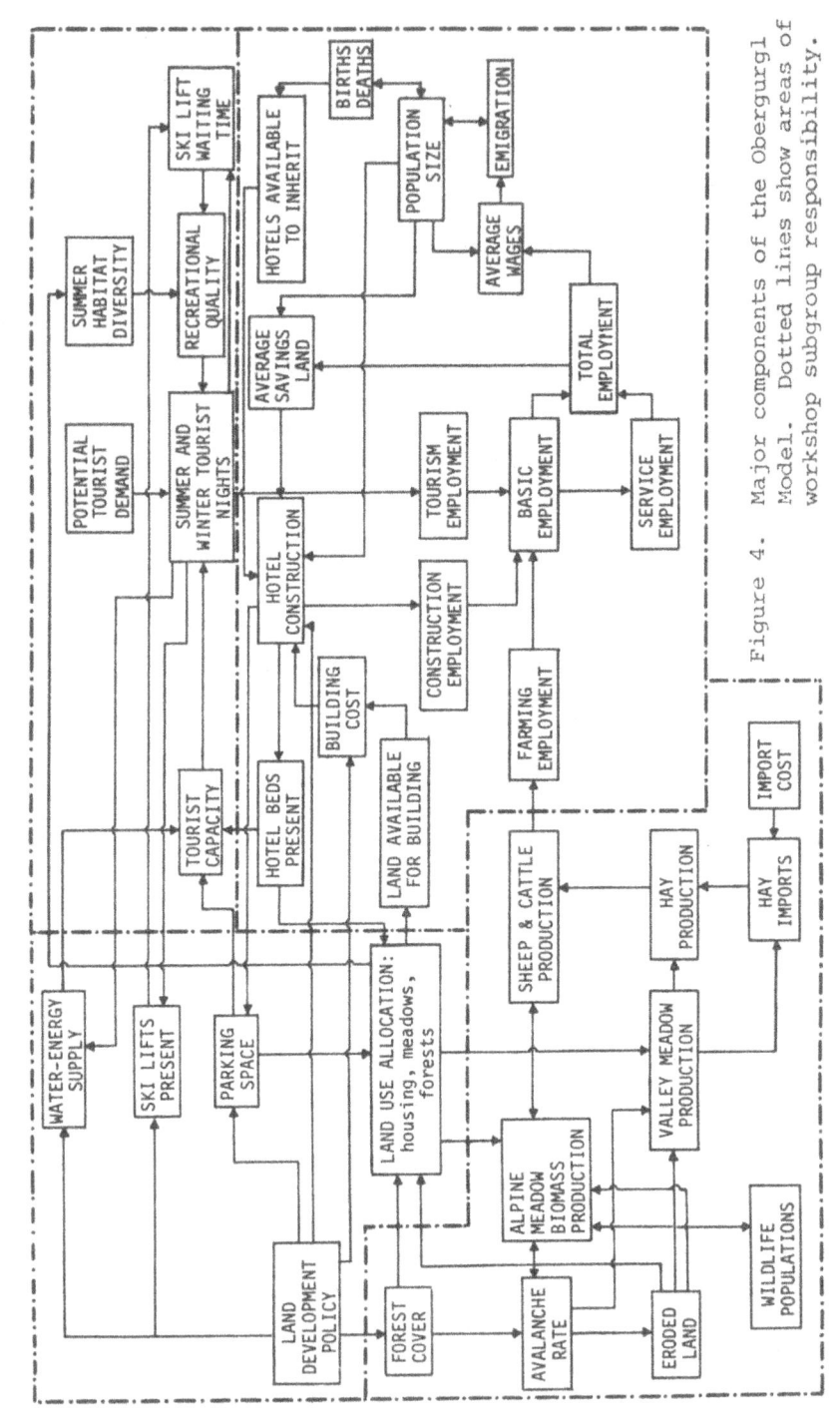

Figure 4. Major components of the Obergurgl Model. Dotted lines show areas of workshop subgroup responsibility.

throughout Europe), predicts that stabilization of demand
will not immediately stop growth. It is assumed here that
over-investment in hotels will occur until no owners are
doing very well. On the positive side, a continued demand
crisis should help to spread the inevitable emigration pulse
over a longer period of time, so that widespread social
dissatisfaction would not develop all at once.

Other scenarios included assumptions of a government
subsidy to assist the young to build hotels; government taxes
to make building more difficult; limitations of services (ski
lifts, water) provided for tourists; land zoning; and so on.

From the variety of scenarios that were tried, some most
likely and some most extreme predictions can be drawn:

(1) Even if meadow land for building were not limited,
    the village would probably not grow to more than
    150 hotels (double its present size) by the year
    2000, based on the number of young people who are
    likely to reach the house-building age. The most
    likely prediction is 80-90 hotels present when the
    village reaches its safe land limits in about 20
    years.

(2) Hotel building will not significantly alter the
    amount of valley grazing meadow in the near
    future; only about 20% more of this land is ever
    likely to be developed.

(3) With no land limits, the local population could
    reach 700 persons by the year 2000, with a tourist
    use of about 600,000 nights per year. The most
    likely estimate for population is that equilibrium
    will be reached near the turn of the century, at
    500-600 persons with a tourist use of about
    350,000 nights per year. The most likely
    population growth rate for the next decade or two
    is 2.6% per year, considering the increases that
    are likely in emigration rates.

The ecological implications of these predictions were
not made clear by the modelling work, since the ecological
data base is still very poor. Present recreational use may
already be more than the sensitive alpine meadows can tolerate;
doubling of recreational use is probable and may be disastrous.

Recommended Research Projects

The following are recommendations for research projects

on Obergurgl, listed in order of rank:

(1)   Sociology of villagers in relation to attitudes about land ownership, emigration, and economic opportunities.

(2)   Perception of environmental quality by villagers and by tourists, initially by means of photographic scenarios of future possibilities.

(3)   Basic mapping of ecological conditions in the area, especially in relation to ski development and soil erosion.

(4)   Determination of primary production of pastures and alpine meadows in relation to grazing by wild and domestic animals.

(5)   Projection of potential recreational demands in relation to changing transportation systems and public attitudes across Europe.

(6)   Continued "policy analysis" of alternative development schemes and research priorities.

(7)   Experimental ecological studies involving manipulation of grazing patterns, trampling of meadows by people, and construction activities.

(8)   Economic analysis of the village in terms of employment structure, savings patterns, and cost problems in hotel construction.

In retrospect, it appears that the model described in this section can, after some relatively minor refinement, provide a solid basis for predictions about the human aspects of environmental change in Obergurgl. It remains for future modelling work to develop the ecological side of the story more fully, so a truly balanced picture of the overall system can eventually emerge.

Conclusions

It is obvious that the great contrasts in land ownership, federal and state laws, and behavioral patterns, to mention only a few, between the Austrian Alps and other mountain areas suffering from tourist impact, will necessitate significant modifications and adaptations if the Obergurgl Model is to be applied extensively. In addition, alternate approaches may be used parallel with, or instead of, replications of

the model. Nevertheless, it is maintained that the Obergurgl
Model should facilitate comparative international mountain
research. In many ways its most important function is less
of an explanatory device and more of a catalyst, or a set of
more or less common operational principles. A major recommend-
ation, therefore, is to seek to replicate the model by using
as examples other areas. For further details on the model the
reader is referred to *The Obergurgl Model* (9).

The Colorado Rocky Mountains: Legislative Response to Construction in Hazardous Areas

The mountain areas of Colorado have experienced accelerating pressures from rapid development of the recreation
industry, principally the construction of second homes and
the development of winter sports. The population explosion
along the Front Range urban corridor since 1965 has induced
the completion of the Eisenhower Tunnel under the Continental
Divide, bringing large sections of the mountain ski areas
within 2 to 3 hours driving time of Denver; the twinning of
the auto-freeway (I-70); and the creation of a new type of
boom town, the ski resort, as exemplified by Vail and Aspen.
The inflow of population has placed large numbers of people
with little or no mountain experience in high mountain terrain.
Land values exceed $70,000 per acre in some of the more
attractive sites, and land speculation has developed rapidly.
Since only a limited amount of land is suitable for construction,
the inevitable result--a combination of speculation, ignorance,
and the very speed of the development itself--has been land
sales, and actual construction, in areas subject to a variety
of natural hazards, including avalanche, landslide, debris
flow, rockfall, and mountain flood.

Initial avalanche and debris flow hazard mapping in the
vicinity of Vail and the northern San Juan Mountains by staff
of the University of Colorado, Institute of Arctic and Alpine
Research (10, 11, 12), and a vigorous public campaign by the
Colorado Geological Survey, among others, brought before the
public the rapidly growing danger of uncontrolled construction
in mountain areas. This produced a definite legislative
response with the passage of Colorado State House Bill 1041
which, in part, requires each county to prepare maps of land
subject to natural hazards. This in turn permits the county
planners to require the applicant for a construction permit to
produce acceptable evidence that the proposed construction
site is free of hazard. Some counties produced their own
avalanche zoning ordinances, the town of Vail changed its
building code and annexed a large tract of neighboring land
(13), and an extensive move began to develop a scheme for
systematic mapping of natural hazards at a scale of 1:24,000

(12). In the process of all this activity, some construction companies have gone bankrupt, and a half-complete hotel has been dismantled and its site returned to park-land. The process so far is in its early stages and has served to emphasize how little we know about precise definition of degree of natural hazard, and even our inadequacy in being able to produce meaningful hazard maps that can be used effectively by the planner. At least the first important legislative step has been taken giving a good measure of constraint upon uncontrolled development and allowing a breathing space so that applied and basic research can begin to catch up and match the mounting demands.

Discussion

Assessment of our actual response to the problems facing these three mountain areas, of course, contrast sharply. For the tropical Andes the situation is very complex and more nearly comparable to the Himalaya; more detailed discussion has, therefore, been devoted to this area than to the Alps and the Rocky Mountains. First, large amounts of international money will be needed, both for the development of the research program as well as for the initiation of applied pilot projects. In addition, close collaboration between local scientists and extra-regional specialists will be necessary, both for carrying out the research and for training young scientists to bridge the gap between the natural and human sciences. It has become a principal aim of the Unesco Man and the Biosphere (MAB) Programme, in collaboration with the United Nations Environment Programme (UNEP), to assist the group of Andean nations involved. A first step will be the publication of a state-of-knowledge report on the Andean System as a whole so that the precise nature and magnitude of the problems can be more clearly defined, existing knowledge identified, and critical gaps in knowledge delineated; finally, research strategies will be developed. Obergurgl is a useful, if miniscule, example of the Alpine situation at large. Here Austrian scientists, federal and local government officials, and local hoteliers and farmers are working together to ensure amelioration of current destructive trends. Similar studies are being developed in other areas of the Alps, including the Grossglockner massif, Grindelwald, Aletschwald, and Briançonnais. In these cases, including Obergurgl, the involvement of Unesco MAB has been the provision of leadership, incentive, and moral support, rather than direct financial assistance, but a delicate balance must be kept and research and planning momentum maintained. In this respect further support for the development of the center for mountain research and documentation located at Obergurgl is needed (Gesellschaft für Gebirgsforschung und Dokumentation,

Obergurgl). A resolution for development of such an international system was acclaimed during the Munich Workshop on
Mountain Environments held in December 1974. Finally, the
example from the Colorado Rocky Mountains is a much more
local affair. In this instance a variety of local groups,
including State government scientists, university and private
groups, and individuals, were able to recognize a growing
problem and to produce a sufficiently effective public
campaign to induce passage of new legislation. This provides
a good first step in arresting the more serious aspects of
construction in mountain areas threatened by a variety of
natural hazards. Nevertheless, a large segment of the public
and scientific pressure, essential for this partial success,
received inspiration from the Unesco Man and the Biosphere
Programme.

What practical lessons can we draw from these examples
of problems, partial solutions, and initial responses, from
other parts of the world that can assist the heavy tasks
facing us in the Himalaya? First, I think that the Unesco
MAB reports provide a source of outline research design and
inspiration. Also, while the MAB-6 regional meeting, held in
Kathmandu in September 1975, certainly did not solve any
problems, it assisted with the further identification of
research needs, provided a better outline of methods, and
brought together specialists from many countries with a
common interest. It also identified the pressing need for
establishment of a mountain research institute within the
Himalayan region.

It is perhaps presumptuous for someone with no Himalayan
experience to propose a research approach and a list of
requirements. Nevertheless, some suggestions will be made in
all humility, and, as an excuse for presumption, I believe
that there is some common ground among all the world's
mountain systems, that experience can be shared, and that
serious concern is more important than the fear of reproach
from one's more learned and experienced colleagues. Above
all, the need for an international mountain research community
is emphasized. The total available experience is so sadly
inadequate for the tasks ahead, that we must share every
idea, every practical result, and every item of moral and
spiritual support. Thus, any attempt to improve international
communications among mountain scientists and planners is
worthy of consideration.

In a more direct sense, it is worth emphasizing that
many of the practical problems involved in our response to
social and environmental deterioration in the Andes also face
us in the Himalaya. Thus, the magnitude and complexity of

the problem and the fragmented and inadequate data base cannot be adjusted all at once. It would seem appropriate, therefore, to identify suitable mountain transects so that cross sections of the major east-west divisions of the Himalayan system are represented, from the lowest altitudinal belt to the highest. Then those should be used as the bases for inventory, research undertakings, and development of applied pilot projects of a highly practical nature. A system of biosphere reserves and national parks could be fitted into such a scheme with great effect, and further support given for maintaining the gene pool, such as assistance, for instance, for the impressive work being performed by Dr. Dilip Dey at the Himalayan Zoological Park, Darjeeling, in breeding musk-deer and other threatened mountain species. Furthermore, this approach should be used to provide an incentive for collection and analysis of all existing data, hopefully to be centralized in a mountain research institute with its major task being one of service to the entire region, regardless of international frontiers. Within this framework, four interrelated planning processes are involved: inventory, basic research, application in the form of pilot projects, and long-range political decision making. It must be emphasized that these four processes are parts of a continuum, all of equal importance, and that they merge into one another, since inventory for its own sake, for instance, is of little value. Thus, some forms of inventory would automatically include compilation of maps of actual vegetation, soils, bedrock, surficial materials, climatic elements, land occupation and practices, stream flow, silt content, extent of glaciers and snowline, talus slopes, areas most susceptible to landslides, and so on. Much of this type of inventory is already available for certain areas and in certain forms, but it must be collated and standardized. This material can then be used for deriving another series of maps, such as assumed potential natural vegetation, slope instability, and land capability.

When the work progresses into the basic research phase, human biology must be carefully integrated as has been proposed in the Andean strategy. An extremely interesting and influential process has been *shifting cultivation* (17). There is hardly any region in the Himalaya where shifting cultivation has not been practiced in the past or is not being practiced at present. This process demands careful analysis. Tribal *genre de vie*, environmental degradation, and government laws combine to produce a very complex pattern for analysis. Those laws and value systems which influence the relationship between man and nature for promotion of a higher level of efficiency should be formulated and analyzed, rationalized and implemented. Very little is known about this aspect of the Himalaya; thus a very worthwhile project would be a study of the geoecological implications of laws relating to use of land-wildlife, forests, and agriculture. Then there is a further element that I would like to stress. I think it very important that a serious effort be made to assess the actual impact of post-1950 Man on the Himalayan ecosystems, and I would like to illustrate this by a short digression since it has been common practice to refer increased soil erosion in mountain areas to the accelerating impact of human activites, and this carries with it the danger of over-simplification. The argument proceeds in the following manner: rapid increase in mountain populations since the end of World War II has forced the conversion of forested hill-slopes into grazing and arable land; with continued dependency on local timber for fuel and construction needs, all forested area within easy walking-gathering distance of many settlements has been depleted; a third sector of this vicious circle is that animal excrement, the basic source of fertilizer, is then used increasingly for fuel, which in turn reduces soil fertility and so makes the hillsides even more susceptible to erosion; the final sector, closing the circle, is that soil erosion causes further loss of good agricultural land which in turn induces additional deforestation for planting crops.

In mountain lands, high altitudes and steep slopes combine to form, in a geomorphic sense, a high energy environment as outlined in the introduction. The already great potential for very effective geomorphic activity, including a range of processes from soil creep, through landslides, to debris flows and associated catastrophic events, already exists in nature. The human impacts, setting in motion the vicious circle outlined above, when imposed upon this natural high energy system, may augment mass movements to a degree several orders of magnitude greater than those occurring under natural conditions.

While it is eminently correct to apply this reasoning to one of the most serious complex of causes that drives the alarming loss of fertile agricultural land and forest on the mountain slopes of the Himalaya, the question of "how much" must be answered if intelligent *long-term* responses are to be developed. Thus, it is not sufficient to argue that soil losses in the mountains, with their attendant and perhaps more serious downstream effects of flooding and siltation on the Indo-Gangetic Plain, are enormously higher, for instance, than prior to World War II: we need to know by how much. Without a quantitative answer *short-term* policies, themselves of a very expensive nature, could be put into effect without any certain prospects for being viable in the long-term. This position can be supported by recently accumulating information in the broad fields of climatic geomorphology and paleoecology. A gross simplification is to state that climate changes through time and that catastrophic geomorphic events are often related to weather phenomena themselves of a catastrophic nature with a long recurrence interval.

In recent years geomorphologists have been directing increasing attention to the catastrophic event--landslide, torrent, debris flow--that has a very long recurrence interval and hence is extremely difficult, if not impossible, to predict. Because of this difficulty in prediction, of course, it is not yet possible to provide quantitative assessments of the amount of work (movement of unit mass through unit vertical distance in unit time) accomplished by such catastrophic events, nor to fully understand their importance in relation to the work accomplished by slow-moving, continuous processes, such as soil creep and frost creep. Nevertheless, realization is growing that the rare, catastrophic event will account for a high proportion of the total energy expended in mountain slope development. A recent issue of the quarterly journal *Geografisker Annaler* (14) has been devoted entirely to the question of rapid mass movement in different climates. It includes papers that provide case studies supporting the general principle outlined above, and indicates its importance in such contrasting climatic regions as Hong Kong, New Zealand, Central Japan, New South Wales, the Carpathians, Appalachia, and Swedish Lappland.

Let us now consider a specific example in the eastern Himalaya near Darjeeling. In 1968, at the end of the monsoon season when mountain soils would have been well-saturated, 64 cm of rain fell during a three-day period. In the closing hours of the rainstorm some thousands of debris flows were released, directly or indirectly causing the loss of at least 20,000 lives and countless damage to roads, railways, houses, and crops. A measure of the intensity of the event is that the 56 km mountain road that connects Darjeeling with Siliguri on the Ganges Plain, was cut in 92 places. As a very much concerned visitor to the area within four weeks of the occurrence of the disaster, I was told that the authorities had assumed that the basic cause was extensive deforestation, and that the solution, therefore, was a program of reafforestation. The climatological officer in Darjeeling informed me that such a quantity of rain, at such a critical time of year, had only occurred once before since records were first maintained in the latter part of the 19th century. This may mean that we are dealing with a catastrophic event that has a recurrence interval of between 100 and 500 years. In practical terms this would indicate that deforestation may well have heightened the magnitude of the event, but it may not have

been the primary cause. Thus, we must assume that a large
number of debris flows would have occurred even if the
mountain slopes had been protected by a cover of untouched
forest. It must also be pointed out, however, that a significant number of debris flows began at the bends in the mountain
roads and trails, so that these ephemeral and artificial
stream beds contributed significantly to the total number of
individual releases.

The lesson to be learned from the above example is that
catastrophic events are complex in nature and that their magnitude and frequency will be influenced by a variety of human
and natural conditions. A case study of landslide activity in
the Razorback area of New South Wales by Blong and Dunkerly in
the issue of *Geografisker Annaler* referred to above (15), even
raises the possibility that there deforestation by man may
have reduced the susceptibility of the area to landsliding.
These authors go on to a slightly more cautious conclusion:

> During the last few decades it has become fashionable
> to promulgate the influence of human activity on
> landslide incidence. The literature on Razorback
> proves no exception although Hanlon (1958), Cambage
> (1924) and Dunkerley (1974) indicate either that
> landslides would have occurred anyway or that some
> evidence suggests that landsliding predates clearing
> of the natural vegetation. (15, p. 145)

I certainly do not wish to infer any degree of similarity
between New South Wales and the Himalaya. The purpose of
introducing the two examples is to argue that the problem in
general is highly complex and that simple assumptions and
simple responses are potentially dangerous. The complexities
of the natural hazard question certainly cannot be overempha-
sized. In some areas, such as the Simla State region, lithology
and structure render landsliding inevitable regardless of
land-use practices, so that here the question is one of
matching land-use to a prevailing situation. A valuable
approach would be to undertake the mapping of the distribution
of types of hazard prevalent in different regions, which
should also take into account the recurrence of earthquakes.
It is apparent that the lower Himalaya are radically different
in this respect from the Trans-Himalaya. Such an approach
would then form the basis for development of international
analogs and the formulation of general models. It is upon
this basis that a case can be made for the importance of
assessing the actual impact of human activities on slope
processes in the Himalaya, and that one of the best approaches
is a combination of paleoecological investigations and study
of extant geomorphic processes. However, it is also stressed

that such an approach is not considered as an alternative to a short-term crisis response. If soil erosion, in any area, can be checked, even temporarily, by any means whatsoever, then those means should be put into effect provided that they do not merely achieve a very short-term gain at the cost of a slightly delayed catastrophe.

Paleoecological research, therefore, is conceived as a long-term research goal that should help in the development of long-term applied results. It should also be carried out in conjunction with the collection of all available data of a historic nature on the past occurrence of catastrophic events; ideally this would involve collection of data from historic geomorphic studies, analysis of all forms of available documentation, and application of anthropological and archeological techniques. Stratigraphy, in its broadest sense, should form the basis of any applied paleoecological research program. Thus, it would include palynology, paleontology (study of macrofossils in peat and lake sediments), historical geomorphology, dendrochronology, and archeology, together with the application of climatic transfer functions so that stratigraphic interpretations can be converted into a picture of changing climate through time. Rates of natural siltation of Himalayan lakes could be obtained from the study of deposits of former lakes, dammed, for instance, by landslides and drained catastrophically several thousand years ago.

Radiocarbon dating of upper and lower horizons in such a sequence of lake sediments should permit the calculation of siltation rates, while pollen and macrofossil analysis would indicate the prevailing vegetation--climate types and enable analogs to be developed with comparable situations of today. By selecting former lakes of different ages and in different climatic zones of the Himalayan system, the pool of data and interpretation, of course, could be greatly enlarged.

This, however, is only one of many possible lines of approach. It should provide a better means of estimating the useful life, for instance, of any proposed future reservoir. It would also provide a much fuller understanding of the actual increase in rates of soil erosion due to human impact as compared to naturally high rates characteristic of steep mountain slopes. Similarly, it has been argued that detailed knowledge of climatic change through the last 10,000 years, and especially the last 1000 years will provide a much better basis for estimating magnitude and direction of future change. One very effective method for determining former vegetation covers is through analysis of soil micro- and macrofauna which remain long after their associated vegetation structures have changed (16). It will, of course, be equally important to attempt to separate out anthropogenic processes from the other processes that have occurred and fluctuated over the last several thousand years.

Knowledge of the recurrence interval of specific types of catastrophic event, while difficult to obtain, would also prove of great practical value. In terms of any natural hazard reduction policy development it is desireable to better understand the magnitude of the impact and its frequency of occurrence. This is because defense structures, or strategies, must be designed to a specific probability of danger, defined as the product of individual event magnitude and recurrence interval. With large impact events occurring above a certain recurrence interval (i.e., less frequent), an intelligent decision must be made between enormously expensive response and no defense at all. Similarly, great advantages can accrue from being able to assess the ratio between mass movement induced by human impact and that which would have occurred on the same site prior to the current population explosion. These two problems, therefore, are represented as the core of any program in applied paleoecological research in mountain regions.

Finally, one of the major human-geographical problems of the Himalaya is the large out-migration from the lower socio-economic classes to the Plains in search of jobs, both seasonal and permanent. There is a pressing need for systematic identification of areas experiencing this out-migration and for an understanding of the variables influencing this process and its impacts on both the source and receiving areas. Here the studies in the Andes by Paul Baker and his group (8), should provide useful insights.

In summary, the applied research needed to tackle a problem as immense as environmental and social deterioration in the Himalaya and other mountain areas requires an integrated interdisciplinary and international approach that will study current processes along a series of transects together with examination of changes in process through time. It will be very difficult to mount successfully and will not yield short-term results, but life on this planet is, hopefully, a long-term problem, and long-term programs must be developed. This is not enough, however, since the current rate of despoilation is so severe that a crisis response is also required. Here the role of national governments and international agencies must be coordinated. Immediate engineering, agricultural, social, and reafforestation responses must be taken. Some critical areas can be emphasized. Road construction needs careful attention, for instance, in light of the implications of the Darjeeling floods and debris flows; the

upper timberline is especially vulnerable and all possible
steps should be taken to protect it; some technological
break-through is required to provide an alternate source of
fuel, major engineering works should only be undertaken when,
say, 5% of their budget has been used for environmental and
sociological assessment of their potential impact, and when
their results have been incorporated into design criteria.
Land-use alternatives and their potential social and environmental consequences must be explored.   The list is endless.
The problems are further exaccerbated by the bald realization
that human and environmental scientists are not presently
capable of proposing a series of changes that will lead to
the optimization of human welfare in an environmentally
balanced system.   This point must be made very clear at the
onset of any research program.   In taking a more realistic
and cautious approach, however, it should be possible to set
up guidelines from which the consequences of land-use alternatives could be anticipated, and actions leading to the
degradation of either the environment or the human system
avoided.   Thus, two points of view must be balanced: the
impact of human activities on the environment, and the impact
of environmental processes on individuals and societies.

Conclusion

Assuming that we have all the inventory material and all
the research results needed, the task will have barely begun.
Thus, we will need concerted action by government departments,
universities, and private organizations within individual
countries, and extensive international collaboration.   But
this also will not be enough unless we can achieve the active
and enthusiastic participation of the people who, day-by-day,
year-by-year, use the mountain lands.   This is perhaps the
most difficult phase.   Finally, we cannot await the orderly
completion of all of these tasks.   The battle must be joined
immediately on many fronts: short-term crisis solving, case
studies, small-scale pilot projects (all necessary for various
reasons, and the public relations reason not the least).
Thus, ideally, all these should begin while the more ambitious
research monitoring, and long-range planning, is being
undertaken.

The task may seem so large and complex that we should
all despair.   Such shunning of responsibility is intolerable.
Some of the more serious threats to world civilization and
environment are lethargy, complacency, despair, conviction
that the problems are too great, the time too short, and that
the damage already accomplished is irreversible.   The question
"can the scientist assist in the preservation of the mountains?"
cannot be answered with a definite affirmative; the issue

hangs unresolved, and it is, at least in large part, a political issue. Do we have the collective expertise and determination? The answer to this question is, probably, yes, at least to make a worthy beginning. But the political issue may overpower us without a very large measure of determination. Can individual communities, as well as national governments, and international organizations of many hues, be induced to think and act? We certainly cannot give a resounding yes to this question, neither can we shrink from making the attempt to influence the outcome.

Acknowledgments

In writing this paper I have drawn heavily from my involvement with the Unesco Man and the Biosphere (MAB) Program, Project 6: study of the impact of human activities on mountain and tundra environments, and from the work of the International Geographical Union Commission on Mountain Geoecology. I should like to thank Dr. J.S. Lall, India International Center, New Delhi, for permission to publish this paper, much of which appears in a chapter I wrote for his forthcoming book on Himalayan environmental problems. The section on the Andes has been virtually borrowed from a special report prepared for the United States National MAB Committee (Ives and Stites, Editors, 1975) and written by Paul T. Baker, Brooke Thomas, and Jere D. Haas; likewise the section on the Obergurgl Model has been derived from the same work, and originally from the work of Walter Moser and Himamowa (1974). Studies of natural hazards in Colorado were supported by the United States National Aeronautical and Space Administration (NASA), Office of University Affairs (grant no. NGL-06-003-200 to J.D. Ives as principal investigator). I am indebted to many colleagues for stimulating discussions and encouragement including Herbert Aulitzky, Paul T. Baker, Francesco di Castri, Olivier Dollfus, Herbert Franz, Gisbert Glaser, James M. Harrison, Corneil Jest, Donald King, Ricardo Luti, Bruno Messerli, Walter Moser, A.B. Mukerji, Brooke Thomas, and Sven Zethelius. Time for the writing was provided through the award of a John Simon Guggenheim Memorial Fellowship and a Faculty Research Fellowship from the University of Colorado, held as Guest Professor at the Geographische Institut der Universität Bern.

References

1. J.S. Lall, *The Himalaya--A Study of Change*. Oxford University Press; in press (1978).

2. IGU Commission on Mountain Geoecology Bulletin No. 1. 8 pp. (1977).

3. Unesco, Programme on Man and the Biosphere (MAB), Expert Panel on Project 6: Impact of Human Activities on Mountain Ecosystems, Salzburg, January 29-February 4, 1973, Final Report. *MAB Report* 8, Unesco, Paris: 69 pp. (1973).

4. Unesco, Programme on Man and the Biosphere (MAB), Working Group on Project 6: Impact of Human Activities on Mountain and Tundra Ecosystems, Lillehammer, 20-23 November 1973, Final Report. *MAB Report* 14, Unesco, Paris: 132 pp. (1974).

5. Unesco, Programme on Man and the Biosphere (MAB), Working Group on Project 6: Regional Meeting for the Andes, LaPaz, 9 to 16 June 1974, Final Report. *MAB Report* 23, Unesco, Paris (1974).

6. Unesco, Programme on Man and the Biosphere (MAB), Working Group on Project 6: Regional Meeting for the Mountains of Southern Asia, Kathmandu, Sept. 30-Oct. 4, 1975, Final Rept. *MAB Report* No. 27, Unesco, Paris (1977).

7. J.D. Ives and A. Stites (eds.), Unesco Programme on Man and the Biosphere (MAB) Project 6: Proceedings of the Boulder Workshop, July 1974, Colorado. *INSTAAR Spec. Publ.*: 122 pp. (1975).

8. P.T. Baker and M.A. Little (eds.), *Man in the Andes: A Multidisciplinary Study of High-Altitude Quechua.* US/IBP Synthesis Series I. Dowden, Hutchinson and Ross, Inc., Stroudsburg, Pa.: 482 pp. (1976).

9. Bubu Himamowa, *The Obergurgl Model: a Microcosm of Economic Growth in Relation to Limited Ecological Resources.* Conference Proceedings from Alpine Areas Workshop, Coordinators, H. Franz and C.S. Holling. Laxenburg, International Institute for Applied Systems Analysis: 80 pp. (1974).

10. R.L. Armstrong and J.D. Ives (eds.), Avalanche release and snow characteristics, San Juan Mountains, Colorado. *University of Colorado, Institute of Arctic and Alpine Research, Occas. Pap.* 19: 256 pp. (1976).

11. J.D. Ives, A.I. Mears, P.E. Carrara, and M.J. Bovis, Natural hazards in mountain Colorado. *Ann. Assoc. Amer. Geogr.*, 66(1): 129-144 (1976).

12. J.D. Ives and M.J. Bovis, Natural hazard maps for land-use planning, San Juan Mountains, Colorado, U.S.A. Proceedings of International Geographical Union Commission on Mountain Geoecology Symposium, Caucasus, 1976. *Arct. Alp. Res.*, 10(2), in press (1978).

13. J.D. Ives and P.V. Krebs, Natural hazard research and land-use planning responses in mountainous terrain: the Township of Vail, Colorado Rocky Mountains. Proceedings of International Geographical Union Commission on Mountain Geoecology Symposium, Caucasus, 1976. *Arct. Alp. Res.*, 10(2), in press (1978).

14. A. Rapp and L. Stromquist (eds.), Case studies of rapid mass movements in different climates. *Geogr. Ann.*, 58A(3): 127-200 (1976).

15. R.J. Blong and D.L. Dunkerly, Landslides in the Razorback area, New South Wales, Australia. *Geogr. Ann.*, 58A(3): 139-148 (1976).

16. H. Franz and F. Kral, Pollenanalyse and Radiokarbon-datierung einiger Proben aus dem Kathmandubecken und aus dem Raum von Jumla in Westnepal. *Akad. Wissenschaften, Math.-naturwissenschaften Kl.* Abt. I, 184(1-5) (1975).

17. J.E. Spencer, *Shifting Cultivation in Southeastern Asia*. University of California Press, Berkeley: 255 pp. (1977).

3

Monitoring and Mapping Mountain Environments

Daniel H. Knepper

Introduction

Physical studies of the mountain environment deal with a great variety of natural and man-induced problems. Understanding and finding solutions to these problems is becoming increasingly urgent, as the impact of man's presence in the mountainous regions of the world continues to grow and as major land-use questions are considered by the policy makers. In addition to the time constraints on land-use decisions and their implementation, many of the decisions that must be made are irreversible. Therefore, we must ensure that adequate scientific information is available for the development of sound land-use policy.

Existing scientific data in most areas are generally insufficient for land-use planning, and new research programs must be initiated. This research must be comprehensive, definitive, and completed rapidly. Much of this new research involves mapping the distribution of natural features (vegetation, landforms, surface materials) and monitoring changes in the landscape. Mountain research is inherently slow and expensive, largely because of the logistical problems that must be overcome. Ground-based research programs, therefore, usually cannot accomplish the volume of data collection and interpretation needed for timely input to the land-use decision-making process. Remote-sensing technology is helping to reduce this information gap.

Remote sensing, loosely defined, is the use of electromagnetic radiation (EMR) to obtain information about a target at a distance. Many diverse applications of remote-sensing data, as well as a detailed treatment of remote-sensing theory, data processing and analysis, and systems operations, are discussed by Reeves (1). The purpose of this paper is to outline briefly some characteristics of the more common remote-sensing systems that pertain to various types of mountain problems.

Types of Data

Operational remote sensing is generally restricted to the visible, photographic infrared, thermal infrared, and microwave and radar portions of the electromagnetic spectrum. Remote-sensing at shorter (ultraviolet, x-ray, etc.) and longer (radio) wavelengths is seriously hampered by a number of factors, most notably atmospheric absorption, low natural emissions, and man-made and natural noise. Within the usable portions of the spectrum, a variety of remote-sensing systems have been developed to detect, measure, and record EMR. There are two basic types of remote-sensing systems: passive and active. Passive systems detect and measure EMR naturally emitted from the target materials because they are at temperatures above $0^{o}K$ (thermal infrared and microwave). Active systems measure EMR from an external source that is reflected by the target materials.

Visible and Photographic Infrared

Remote-sensing systems used in the visible and photographic infrared bands are all passive; the sun is the external source of EMR. The camera/film system is the most common. The camera collects and directs the EMR to the film where it is detected and recorded. With the appropriate film and a combination of filters on the camera lens, photographs of selected portions of the visible and photographic infrared bands can be obtained to selectively enhance or subdue certain features (multispectral photography). Prior knowledge of the spectral reflectance characteristics of the target and background materials is, of course, necessary to determine the correct film/filter combinations to use. This knowledge can be acquired through trial-and-error, results of previous research, and selected field investigations. The spatial resolution characteristics of photographs are superior to other forms of remote-sensing data and their capability for stereoscopic analysis is a definite advantage for many applications. However, manipulation of photographic data is limited.

To overcome this limitation, data can be acquired with an optical-mechanical scanning system. A scanner rapidly scans across the terrain perpendicular to the flight path of the aircraft, detecting and recording the reflected EMR along each scan line. As the aircraft moves forward, a series of scan lines are built-up to produce an image of the remotely sensed terrain. These data can be photographically recorded on film or recorded on magnetic tape for later computer processing and enhancement. For many applications, the poorer spatial resolution of scanner images is more than compensated for by the increased spectral resolution of scanning systems (the ability to detect small differences in spectral radiance) and the compatibility of the raw data with computer processing. Furthermore, the scanning systems can simultaneously acquire imagery from many narrow bands within the visible and photographic infrared spectrum, all in perfect registration, thus reducing the need for prior spectral reflectance information.

Remote-sensing data from the visible spectrum is probably the easiest to interpret, since our eyes are also tuned to visible light. Color photographs depicting terrain in familiar colors, shapes, and textures are generally the overall most useful type of data, although the unconventional photographs and images are equally powerful in the hands of a skilled interpreter. Photographs and images using the photographic infrared portion of the spectrum, however, allow us to see what our eyes cannot detect, making interpretation less straight-forward. Two factors make the photographic infrared spectrum especially important. First, the spectral reflectance of vegetation in the photographic infrared is extremely sensitive to amount, type, maturity, and growth vigor. Plants with a substantial mesophyll layer in their leaves, such as most deciduous trees and well-watered grasses, are highly reflective in the photographic infrared. However, as this type of vegetation becomes stressed, photographic infrared reflectance decreases sharply, and this can be detected on photographic infrared photos or images long before the effects are apparent in the visible spectrum. Coniferous trees and other less-lush vegetation forms are only moderately reflective in the photographic infrared when they are healthy, and a similar decrease in photographic infrared reflectance occurs as they become stressed. Since many mountain research problems deal directly or indirectly with vegetation, there are many ways to apply the above ideas using remote sensing data. For example, the Institute of Arctic and Alpine Research (INSTAAR), under a grant from the National Aeronautics and Space Administration (NGL 06-003-200), has successfully used color infrared photography (discussed below) to map a variety of natural hazards in the alpine terrain of the San Juan Mountains of southwestern Colorado, including snow avalanche paths cut through coniferous forest and several ages of aspen revegetation in certain avalanche paths, areas subject to mass failure defined by disturbed vegetation, and swamps and bogs.

Second, water largely absorbs the photographic infrared radiation from the sun. On photographs and images from the photographic infrared band, water bodies stand out in sharp contrast with the surrounding terrain and are relatively easy to identify. Furthermore, the boundaries between water and shore are very sharp, so that repetitive photography is an effective tool for monitoring fluctuations in water level.

Color infrared (false color) photography is a common form of remote sensing data using the photographic infrared portion of the spectrum, as well as the green and red bands of the visible spectrum. Photographic infrared radiation exposes the red layer of the film so that objects that reflect large amounts of photographic infrared energy appear in various shades of pink and red. Since the human eye can distinguish many more color tones than shades of gray, the color representation enhances the ability to distinguish subtle variation in the spectral radiance of the surface materials, especially vegetation.

Thermal Infrared

The thermal infrared spectrum includes EMR from about 2 μm to 1000 μm in wavelength. However, atmospheric absorption and generally low levels of natural emissions restrict remote sensing to two bands within the full infrared spectrum: 3 to 5 μm and 8 to 14 μm. These bands are both atmospheric windows, and the 8-to-14 μm band includes the peak portion of the earth's thermal emission curve. Remote sensing in the thermal infrared band is usually done with an optical-mechanical scanner similar to that described for visible and photographic infrared sensing but with detectors sensitive only to the usable thermal infrared bands. The detectors sense the radiant energy in the infrared that is emitted by all objects at a temperature above $0°K$. The amount of EMR emitted by an object is a function of its temperature and thermal properties. If the terrain has come to thermal equilibrium, as it tends to do just before sunrise (all objects are the same temperature), then differences in radiated thermal infrared energy must be due to differing thermal properties (i.e., different materials). This principle can be used to effectively map materials at the earth's surface and may provide information that is not apparent on conventional aerial photos. Alternatively, compositionally homogeneous terrain can be monitored for temperature variations accurately and rapidly. This may be particularly applicable to monitoring thermal pollution and exploring for shoreline hot and cold springs. The presence of water, because of its very high heat capacity, is an important factor in determining the thermal properties of soils. Slight variations in soil moisture content alters the thermal properties enough to allow detection and mapping of relative subtle soil-moisture differences in otherwise homogeneous soils. This fact could be used as a basis for using thermal infrared imagery to rapidly and accurately delineate areas of potential instability caused by natural or man-induced increases in soil moisture.

Thermal infrared imaging is effected by a variety of factors including topography, wind, humidity, and time of day. For qualitative applications of the imagery, only the micro-meteorological factors need to be monitored at several sites during an overflight. However, if semiquantitative or quantiative analysis is anticipated, field parties must also monitor the surface temperatures of target and background materials during the overflight. Interpretation of thermal infrared imagery is also more complicated than aerial photo analysis since the phenomena depicted on the image is totally outside our everyday experience. Therefore, thermal infrared imagery is not recommended for a general purpose tool, but there are many specific applications for which thermal infrared imagery can provide information not contained in any form of photography. For example, bedrock faults covered by a veneer of soil or colluvium often influence the soil moisture along the fault trace, making fault detection and mapping possible using thermal infrared imagery.

Microwave and Radar

Microwave EMR occupies that portion of the electromagnetic spectrum between the thermal infrared band and the radio band, and has characteristics of both. Passive microwave remote sensing is similar to thermal infrared sensing because the source of naturally emitted microwave energy is also thermal emission, controlled by the temperature and thermal properties of the material. However, the levels of naturally emitted microwave radiation are very low compared to the thermal infrared, and sophisticated electronic detectors (antennas and radio receivers) are necessary for microwave sensing. Research has shown that there is a potential for using microwave sensors for effectively observing a limited number of phenomena over the land surface, most notably subtle variations in soil moisture. Microwave remote-sensing, however, is still in the experimental stage of development.

Active microwave remote sensing has been more effectively applied to a variety of earth resources-oriented research problems, and side-looking airborne radar (SLAR) is the most common active microwave sensing system. SLAR systems generate small-scale images of relatively large areas by rapidly beaming microwave energy, generated by the system, toward the earth's surface and recording the returned energy as a function of time (near areas return the energy before far areas). The microwave beam illuminates a relatively narrow swath of terrain to the side of the aircraft during each pulse of transmitted energy, and records the radar echo from that pulse before transmitting the next pulse. During the time between pulses, the aircraft moves ahead slightly, so an image composed of successive scans is slowly built. The dominant feature recorded on radar imagery is macro- and microrelief. Some discrimination of surface materials is possible because of differing reflectance properties at the microwave wavelengths. SLAR is a relatively low-resolution remote-sensing system, but it can accomplish some tasks that other remote-sensing systems cannot. Because of the long wavelength of the microwave energy, SLAR can penetrate clouds and, in some cases, precipitation, enabling the acquisition of images for regional mapping when other systems cannot function. Since the SLAR systems are active, they can produce images any time of day or night. SLAR is not a general-purpose remote-sensing tool and is only for specific applications or special conditions.

LANDSAT Satellite System

The launch of LANDSAT-1 on 23 July 1972 began a new era in the application of remote-sensing technology to pressing environmental and natural resources problems. The overwhelming success of LANDSAT-1 prompted the construction and launch of LANDSAT-2 on 22 January 1975; together these satellite-borne remote-sensing systems have produced imagery of much of the earth's surface.

These satellites carry two imaging systems, a multispectral scanner (MSS) and a return beam vidicon (RBV), which can detect, record, and retransmit electronic imagery data to ground receiving stations for processing. For earth resources applications, the MSS has proven the most useful. The MSS simultaneously produces imagery in four discrete bands in the visible and photographic infrared spectra (red, green, and 0.8 to 0.9 μm and 0.9 to 1.1 μm in the photo IR). Each LANDSAT image covers an area 185 x 185 km. The satellites have been placed in orbits that allow the same area to be imaged every 18 days by a single satellite or every 9 days using imagery from both satellites. Hardcopy, photo-like images of each band of each LANDSAT scene, as well as computer-compatible magnetic tapes of the imagery for computer processing, enhancement, and automatic mapping applications, are available from the U.S. Geological Survey, EROS Data Center in Sioux Falls, South Dakota.

In addition to the two imaging systems, the LANDSAT satellites have a third system with applications to monitoring the high-altitude environment. The Data Collection System (DCS) allows remote monitoring stations equipped with a low-powered transmitter and an antenna to monitor environmental parameters--such as wind velocity and direction, precipitation, and temperature--using the LANDSAT system as a radio repeater link to any NASA ground receiving station. Since there is no data storage capability on the satellite, the frequency of valid messages received from a remote data station (Data Collection Platform-DCP) depends on the time the satellite is in line-of-sight to the DCP and a ground receiving station. Within the United States there is a 0.95 probability of receiving at least one valid message every 12 hours; NASA can supply the data from these messages to the users within 24 hours (2, 3). Specific alpine applications of the system are discussed by Barry and Clark (4) and Kahan (5).

A third LANDSAT system will be launched into orbit in the relatively near future. This satellite will have--in addition to the MSS, RBV, and DCS systems--a capability for imaging in the thermal infrared band that could extend the applications of LANDSAT data to earth-resources investigations into a new dimension.

Summary

The increasing intrusion of man into the mountainous regions of the world and the growing concern that this intrusion is occurring recklessly and without forethought has stimulated research to a level never before seen. The demand for scientific information on the high-altitude environment that can be used as a basis for sound land-use decisions has similarly grown, but the traditional data gathering and analysis methods cannot generate the volume of data that is needed.

A variety of remote-sensing data is available to aid the monitoring and mapping of the mountain environment. However, a casual look at the *Manual of Remote Sensing* (1) will demonstrate that each type of data has its own uses and limitations, and these must be understood before remote-sensing techniques can be effectively applied. The increased use of remote-sensing techniques for environmental and natural resources studies has seriously depleted the pool of experienced personnel needed to conduct new investigations. There is, consequently, a pressing need for natural scientists trained in the application of remote-sensing technology. This training should include a rigorous study of basic electro-magnetic theory, remote-sensing systems and sensors, data reduction, data processing, and data analysis, as well as an introduction to the complexities of mission planning.

References

1. R.G. Reeves (ed.), *Manual of Remote Sensing*: American Society of Photogrammetry, Falls Church, Virginia: 2144 pp. (1975).

2. National Aeronautics and Space Administration, *LANDSAT Data Users Handbook*: NASA/Goddard Space Flight Center Document No. 76DS4258, Greenbelt, Maryland (1976).

3. S. Cooper and P.T. Ryan (eds.), *Data Collection System. Earth Resources Technology Satellite-1*. NASA SP-364, Washington, D.C.: 132 pp. (1975).

4. R.G. Barry and J.M. Clark, Evaluation of an ERTS-1 data collection platform installed in the alpine tundra, Colorado. *J. Appl. Meterol.*, 14: 622-626 (1975).

5. A.M. Kahan, Use of the Landsat-2 Data Collection System in the Colorado River Basin Weather Modification Program. Final Report, U.S. Dept. of Interior, Bureau of Reclamation, Denver: 103 pp. (1976).

4

High Altitude Climates

Roger G. Barry

Introduction

This paper examines the atmospheric aspects of high mountains. Mountains and high plateaus cover at least a quarter of the earth's land surface and yet their climates are very little known, particularly in South America and in Asia. Most classifications of climates simply assign highland areas into a residual and undefined category. This survey outlines the broad patterns of temperature and precipitation regimes, the elements that are most commonly measured in mountain areas, and examines the effects of mountain terrain on these and other climatic characteristics.

It is appropriate to begin by considering the question of the representativeness of weather data from mountain areas. Many stations are sited in valleys for reasons of accessibility and practical convenience. Thus, such data represent the localized conditions of this type of sheltered site. Ridge stations are less numerous, although in Europe there are several mountain summit observatories. Even if stations can be installed and maintained in such locations, these data still represent only the other extreme of the range of possible exposure conditions. Data collection platforms that can transmit meterological or hydrological information to satellites for relay to a ground station and subsequent transmission by teletype, or via the mail on punched cards, have proved valuable already in several remote mountain environments (1,2), but the problem of siting still remains. Within the forest zone, stations can usually be located without difficulty. Above the tree line, however, wind conditions can cause serious problems, especially since much of the precipitation occurs as snow. In the Rocky Mountains, for example, it is found to be essential to equip precipitation gauges with shields and to erect a snow fence around the gauge (3,4).

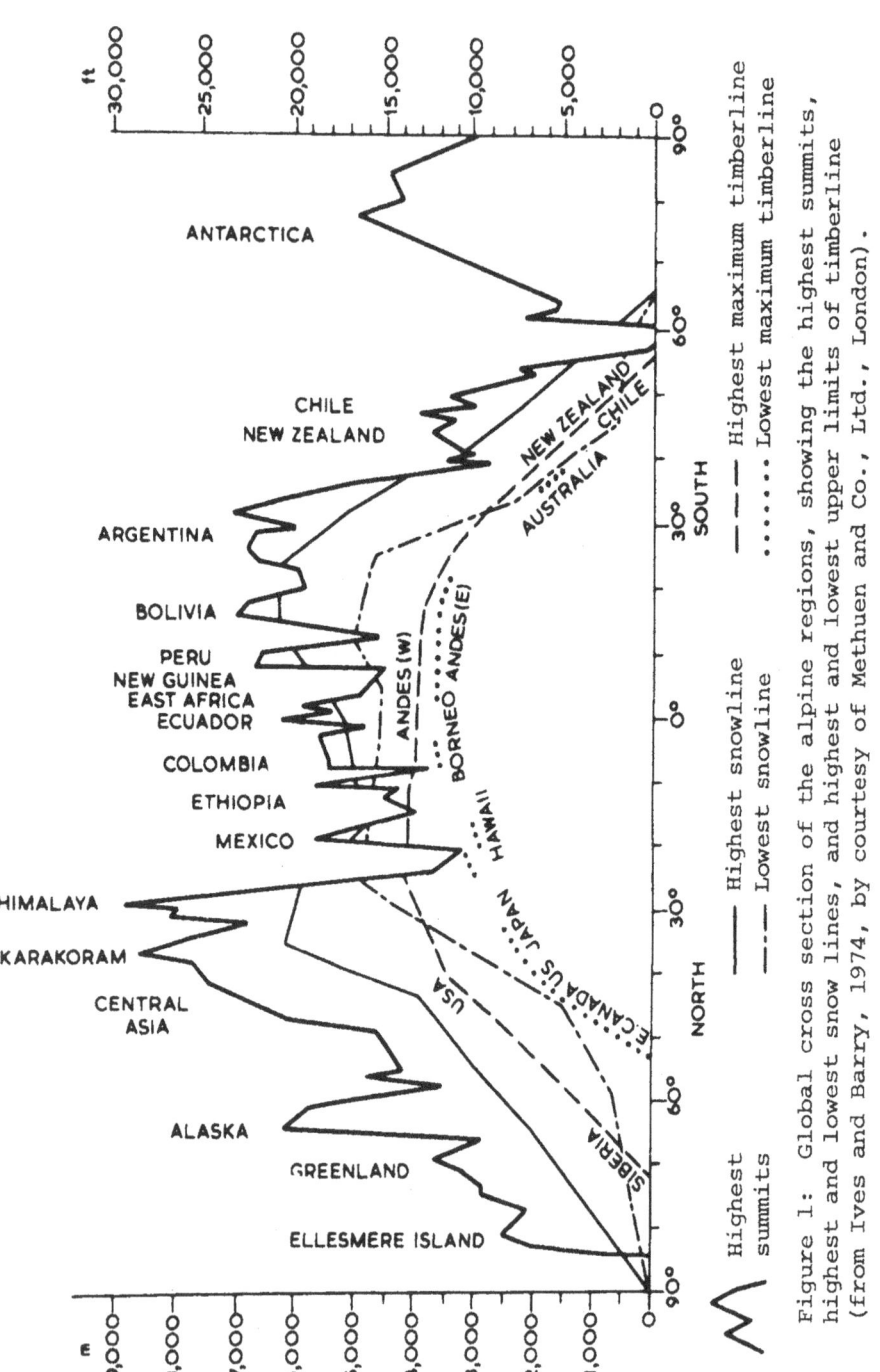

Figure 1: Global cross section of the alpine regions, showing the highest summits, highest and lowest snow lines, and highest and lowest upper limits of timberline (from Ives and Barry, 1974, by courtesy of Methuen and Co., Ltd., London).

Climatic Elements

Examination of a map of global relief shows that there are two distinctive groups of mountain system: the north-south mountain chains of western North America and western South America and the extensive east-west chains of Central Asia with intervening high plateau areas. There are also many lesser ranges and isolated peaks, as well as the high plateau areas of Greenland and Antarctica, although the latter are not considered here. The orientation of the mountain systems with respect to the prevailing wind directions causes their climatic conditions to differ considerably and this point will be referred to again later. The major climatic elements are now examined in turn.

Solar Radiation

There are very few solar radiation data available at high altitude. Measurements made on Niwot Ridge, at 3650 m, in the Front Range, Colorado, during 1973 show that in summer the levels of global solar radiation are closely comparable with those at Barrow, Alaska, where there are 24 hours of daylight (5). There is apparently no increase with altitude in annual global solar radiation on the Front Range (3), as a result of cloud conditions, although certainly during clear sky conditions the radiation amounts at higher altitude are greater than on the plains due to the less dense atmosphere. According to data from the Alps (6), the increase averages 5 to 10% km^{-1}. The effect of increased atmospheric transparency is noticeable in terms of the ultraviolet radiation, with a much greater risk of sunburn at high altitudes. However, the actual altitudinal increase of ultraviolet radiation according to Caldwell (7) may be considerably smaller than that estimated previously (8) since a decrease in the scattered component offsets the increase in the direct component.

Temperature

Figure 1 shows a latitudinal transect of the major mountain systems and indicates the way in which the snow line and the tree line both decline rapidly in altitude in middle latitudes towards the poles. They generally reach their highest elevation in the Subtropics and are generally at a slightly lesser altitude in equatorial latitudes. This is partly a consequence of the radiation and temperature regimes at the Tropics, compared with those along the equator proper, but is also due in part to the opportunities provided by the existence of higher and more extensive mountain systems there than along the equator. This characteristic

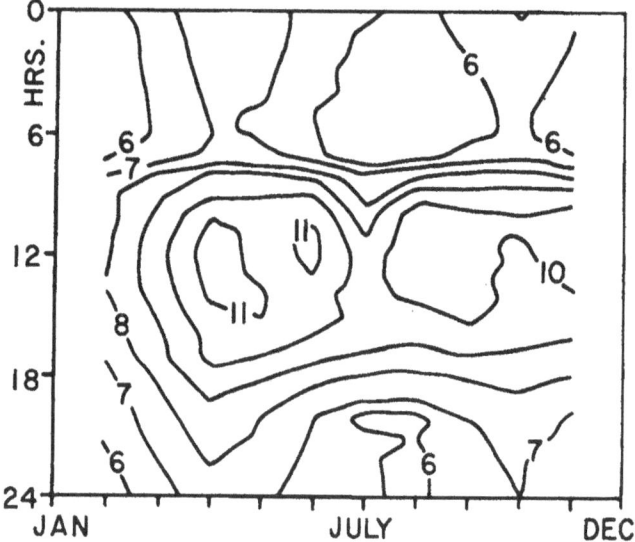

Figure 2.   Thermoisopleth diagram for air temperature (1.5 m)
at Pindaunde, Mt. Wilhelm (3480 m) in 1970 (after
Hnatiuk et al., 1976, by courtesy of the authors
and Australian National University).

is often described in botanical literature as the Massenerhebung (mass-elevation) effect of elevated terrain on summer season isotherms.

In polar regions there is a pronounced seasonal regime of temperature and the virtual absence of any diurnal regime. In contrast, Figure 2 showing temperature conditions at Pindaunde, Mt. Wilhelm, in Papua New Guinea (latitude 5°S, 3480 m) reveals a virtual absence of seasonality, but a relatively well-marked diurnal regime (9). Equatorial high mountains typically experience a nocturnal winter and a daytime summer. Snow which falls overnight on these mountains rarely persists, although there are summits in Irian Jaya, reaching 5000 m, which do carry permanent snow and ice caps (10).

The relationship between latitude and the mean daily temperature range is shown in Figure 3 (11). In low latitudes the mean daily range on summits is of the order of 10°C, decreasing to only 2°C in high latitudes, with a somewhat greater range exhibited at valley stations. It is important to remember that in many parts of the world most high weather stations are in the valleys. Lauscher (11) shows that much of this latitudinal trend can be explained in terms of cloudiness. Thus, in many mountain regions, the cloudiness characteristics are responsible for the magnitude of the diurnal temperature range. Looking at the vertical distribution of temperature range (Figure 4) we note that, in general, in middle latitudes (in the Alps and in the United States, for example) and to lesser extent in Africa (on Mount Kenya and Kilimanjaro) the daily temperature range decreases with elevation. This was first recognized by de Saussure on Mont Blanc in 1796 (12). It is a reflection of the stronger winds at higher altitudes in middle latitudes, which cause greater mixing of the free atmosphere with the air on the mountain summits. The increasing daily range with altitude shown for the Himalaya reflects the plateau location of most of the high stations used in this particular analysis. In the case of Lhasa, for example, the tendency for a strong diurnal temperature regime is related to the low cloudiness at this high plateau station.

Precipitation

Apart from the latitudinal trend, temperature is not a primary feature in terms of differentiating between different mountain systems. Precipitation, on the other hand, is useful in this respect, but only limited data exist for this purpose at the present time. An attempt to categorize the vertical profiles of precipitation against altitude has

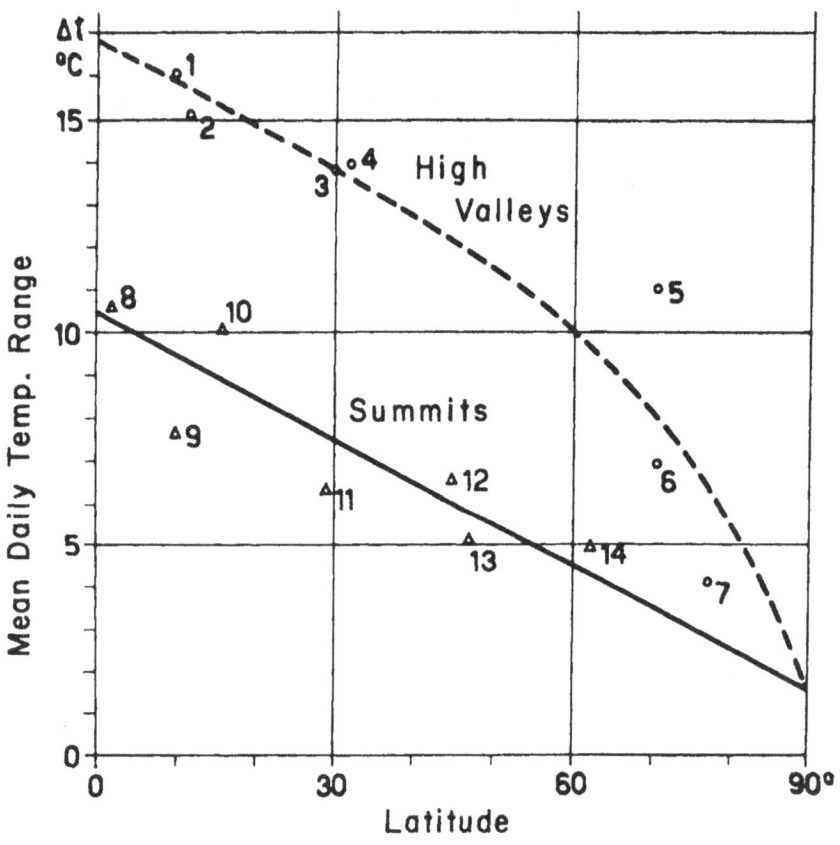

Figure 3: Mean daily temperature range versus latitude for a
number of high valley and summit stations (after
Lauscher, 1966, by courtesy of the author and the
editor of the Jahresbericht Sonnblick-Vereines).

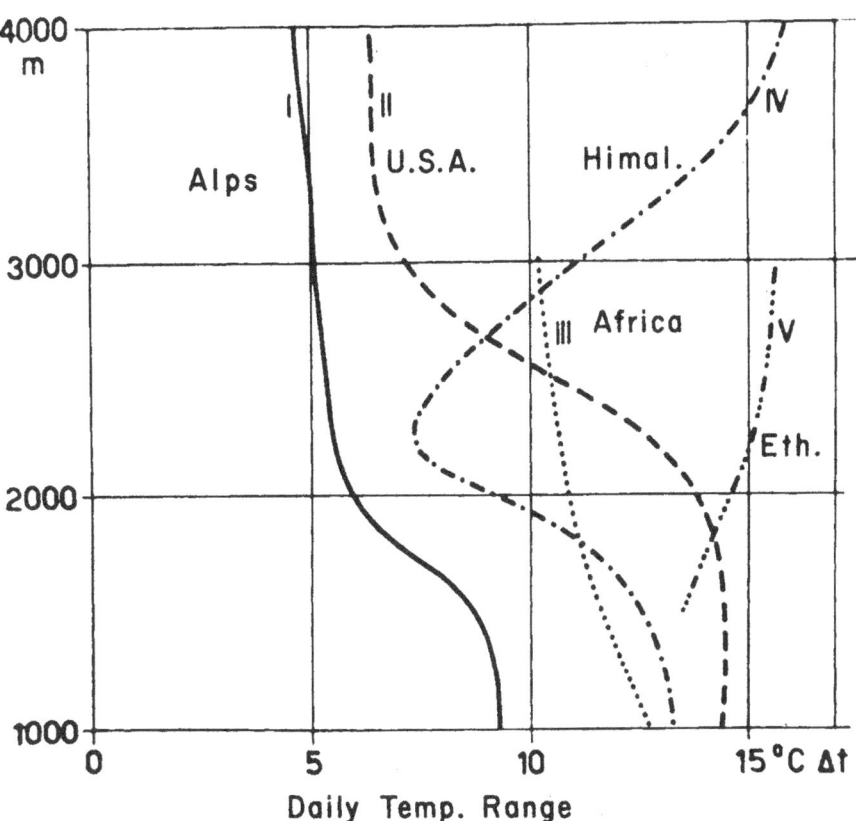

Figure 4: Mean daily temperature range versus altitude in the Alps (i), western U.S.A. (ii), east Africa (iii), Himalayas (iv), and Ethiopian highlands (v). (After Lauscher, 1966, by courtesy of the author and the editor of the Jahresbericht Sonnblick-Vereines.)

recently been carried out by Lauscher in Figure 5 (13).
His schematic diagram shows that, in equatorial latitudes,
there is a general decrease of precipitation with elevation
from the lowest stations. This is also apparent at polar
stations open to maritime influences, although to a much
weaker degree. In the tropics, there is a well-known maximum
at about 1000 m, whereas in middle latitudes there appears
to be an increase up to the highest station levels at about
3000 m. Some of these patterns are now examined in more
detail.

A cross section of height and precipitation from the
south coast of Papua New Guinea to Madang (Figure 6),
indicates that rainfall maxima occur on the lower slopes
between about 800 and 1000 m. On the southern side, this
is due to the forced ascent of the southeast trade wind
flow, whereas it is caused by monsoon westerly flow on the
northern side. Over the central mountains there is a
marked minimum of annual precipitation. In the Tropics,
there is a clear maximum at about 1000 m with a sharp
decrease above. This tropical pattern relates primarily to
the structure of trade wind air with moisture concentrated
in the lowest levels, often capped by an inversion, and
convective-type precipitation (14). In the case of middle
latitudes, both the Sierra Nevadas and the Olympic Range
show that the elevation of the maximum amount occurs rather
higher, reaching at least 1500 m in the case of the Sierra
Nevadas and perhaps as high as 2200 m in the case of the
Olympic Range. In this case, however, we run into the
question of the accuracy of precipitation determinations in
the mountains where a high proportion of the precipitation
falls as snow. Using a two-dimensional numerical model and
simplified topography for the Sierras, Colton (15) has been
able to simulate synoptic cases of precipitation realistically,
for flow normal to the barrier. This type of approach holds
much promise for evaluating the primary controls.

Within the broad zonal patterns, however, important
regional and local variations occur in relation to the
orientation of the barriers with respect to the airflow and
distance to moisture sources (16, 17). The difficulty of
generalization is illustrated by the fact that in Colorado
very different seasonal regimes occur on the west slopes up
to the Continental Divide and on the east slopes (Figure
7). On Niwot Ridge, just east of the Continental Divide,
there is a winter season maximum at the highest station
(3750 m) typical of west-slope sites whereas, on the lower
slopes, there is an early summer maximum even up to 3000 m.
A wide variety of easterly (up-slope) and westerly flow
patterns help to contribute to these regimes. In some

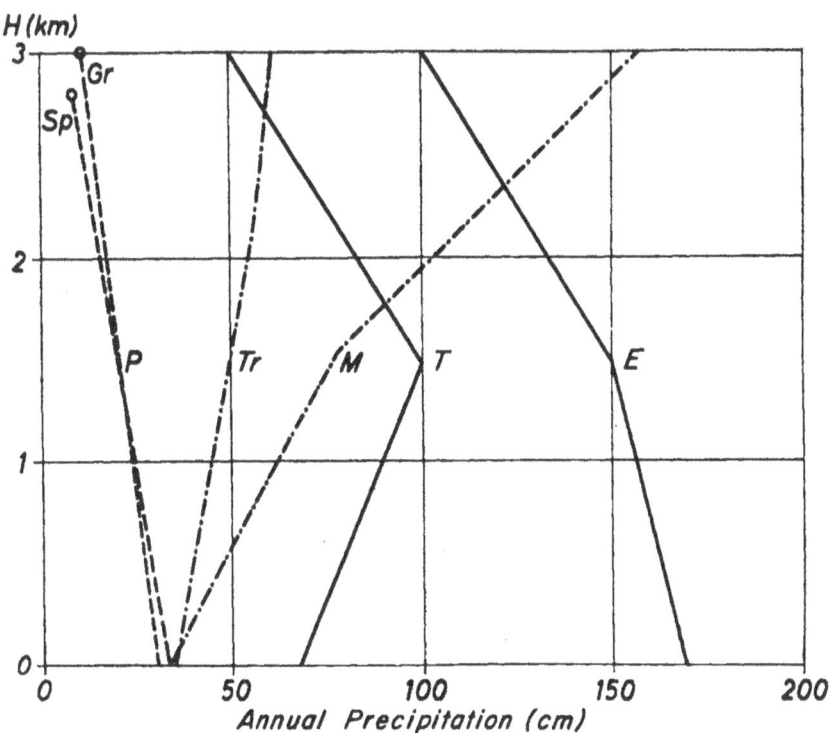

Figure 5: Schematic profiles of mean annual precipitation versus altitude in equatorial climates (E), tropical climates (T), middle latitudes (M), and polar regions (P). Sp denotes Spitzbergen; Gr Greenland; Tr is a transitional pattern between latitudes 30 and 40°N. (After Lauscher, 1976, by courtesy of the author and the editor of Wetter und Leben.)

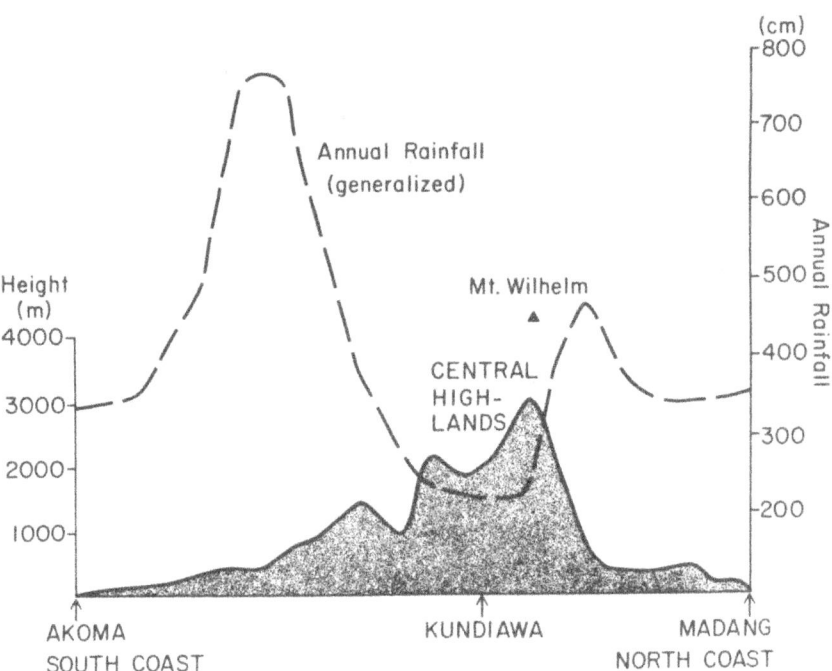

Figure 6: Cross-section of elevation and mean annual
           precipitation from the south coast of Papua New
           Guinea to Madang on the north coast.

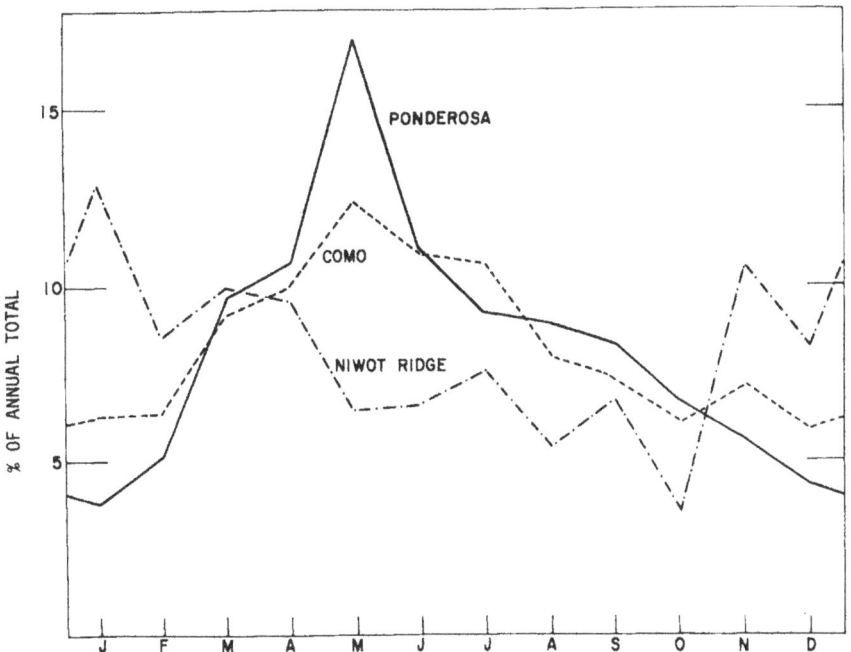

Figure 7: Seasonal precipitation regimes, shown as monthly percentages of the annual total, at stations on the east slope of the Front Range, west of Boulder, Colorado. Ponderosa is at 2195 m, Como at 3048 m and Niwot Ridge at 3750 m. (From Barry, 1973, by courtesy of the Regents of the University of Colorado.)

cases, up-slope flow may reach as high as the Divide and in other cases westerly air extends across the range and may bring precipitation to both sides, although usually in greater amounts on the west slope. Annual precipitation increases with altitude on the east slope to the highest station, but the rate of altitudinal increase is much greater on the west slope (18).

As a result of the problems of precipitation measurement in mountain areas, it is not unusual to find a lack of agreement between hydrological data and precipitation data. Figure 8 from a Norwegian atlas (19) shows that in the Jotunheim-Jostedahl mountains the ratio of the amount of runoff measured in the rivers to the recorded precipitation significantly exceeds unity. Since the precipitation goes partly into evaporation and the balance into runoff this is clearly impossible. What this reflects primarily, is the under-measurement of the true precipitation amount. For the Alps, a recent study by Baumgartner and Reichel (20) indicates that the moisture balance components can be fitted concordantly together. Precipitation there increases with altitude while calculated evaporation decreases slightly. Consequently, there is an altitudinal increase in runoff.

Topographic Effects

The question of topographic effects on weather data concerns both small and large scale influences. The latter, which involve the modification of the global circulation of the atmosphere, may extend far downwind and are not considered here. Within the mountains, a major effect of the terrain is the spatial variability of solar radiation and soil temperature. Local effects are discussed by selected examples. The role of small-scale effects is illustrated by studies at Pindaunde (Mt. Wilhelm) Papua New Guinea. The station is surrounded by high ridges on all sides, except to the southeast. In the early morning there are dense shadows on the northeastern slopes compared with the western slopes of the valley. As a result of the exposure of the east-facing slopes to early morning sunshine, while the west-facing slopes remain in the shade, there is a considerable difference in soil temperatures between these two aspects (Figure 9). This reaches the order of 10°C at 1 cm about 9 to 10 a.m. for measurements taken in September 1975 (21) and is traceable below 7 cm. In the afternoon any advantage in favor of the west-facing slopes is characteristically very weak as a result of the cloudiness that builds up after about 10 to 11 a.m. In order to generalize these conditions, a computer program (22) was used to calculate and map the amount of global solar radiation that would be received at the surface under cloudless skies for the period

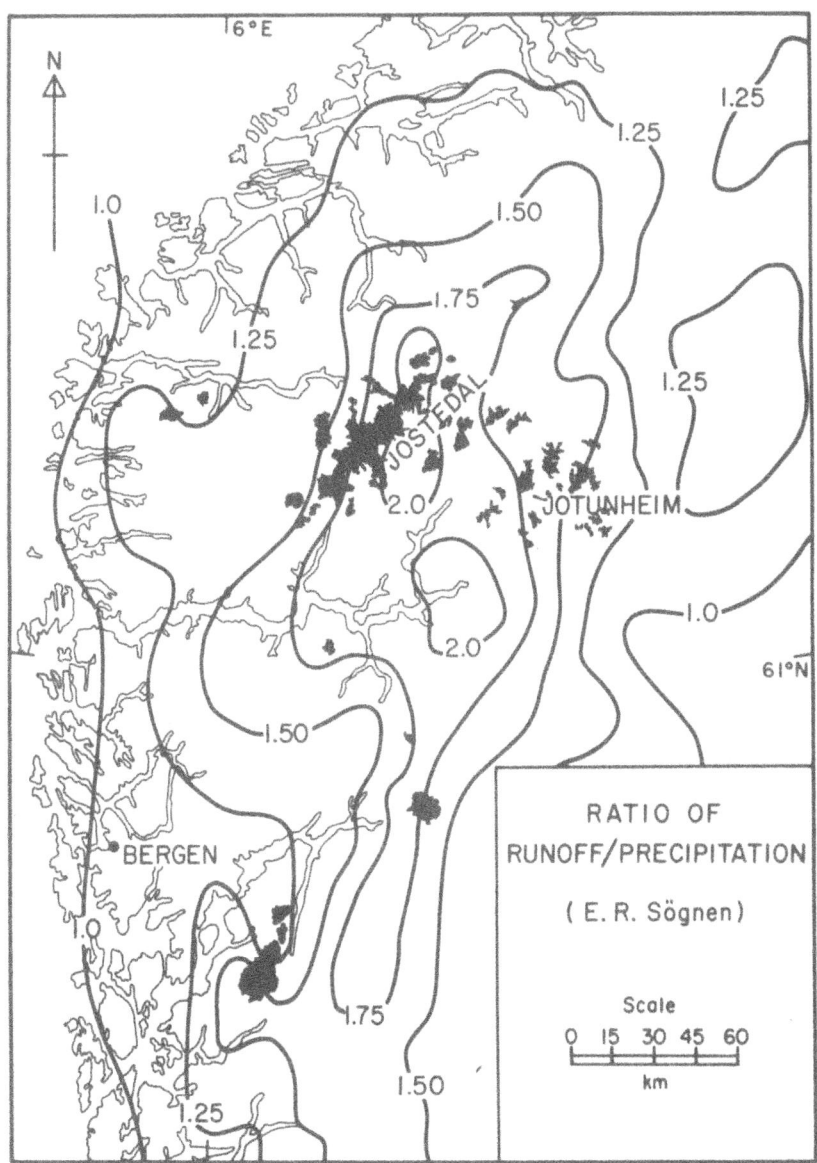

Figure 8: The ratio of measured runoff to precipitation in
central Norway. (After Østrem and Ziegler, 1969,
by courtesy of the authors and the Norwegian Water
Resources and Electricity Board.)

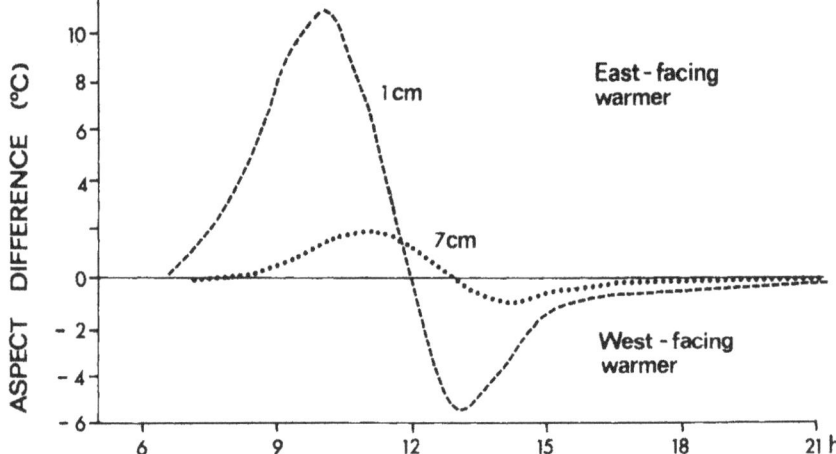

Figure 9: Diurnal differences in soil temperature at 1 cm
and 7 cm beneath tussock grass on east- and
west-facing slopes of about 30° at Pindaunde,
Mt. Wilhelm (3480 m) in September, 1975, on days
with clear sky conditions in the morning hours
(from Barry, in press).

of 6 a.m. to 10 a.m. in September (Figure 10). The east-facing areas (white) have much higher radiation receipts than the west-facing areas (dark) by a factor of 2 to 3 for these hours of the day. This cloudiness and radiation regime is quite characteristic of the dry season (June-August) and transition season (September-October) in that area.

An important parameter strongly influenced by large-scale and also small-scale topography is the wind. Its influence on vegetation near tree line is illustrated by the development of krummholz on many mid-latitude mountains. Snow builds up around the tree islands in winter and affords protection for the limbs growing on the downwind side where the snow dune streams out from the initial obstacle. The upwind side of the trees is presumed to be affected by severe desiccation when the plant roots cannot obtain water due to frozen soil (23).

The small-scale topography determines the collection efficiency of the terrain in terms of its snow cover. Snow banks persist in minor hollows and steps. On a rather larger scale, cirque glaciers and ice patches east of the Continental Divide in the Front Range of Colorado, such as Arapahoe Glacier, appear to be strongly affected by local winds (24). Data collected on the moraine of the Isabelle Glacier show that in winter there is a persistent rotor wind system within the cirque, which is believed to be partially responsible for carrying snow back into the cirque, thereby helping to maintain the glacier (25). Wind conditions are extremely variable over short distances in mountainous terrain. Measurements on exposed knolls on Niwot Ridge in winter 1975-76 compared with the main ridge station about 1-2 km away, show that speeds at the main station exceed 18 m s^{-1} (40 mph) about 3% of the time compared with about 50% on the knolls.

On a larger scale, mountains are responsible for wave motions in the atmosphere, such as those often demonstrated visually by the presence of lenticular clouds. In many of the western mountains of northern and southern America, air flow across the mountains gives rise to Chinook type conditions on the leeward slopes (26). Traditionally, this föhn phenomenon has been explained in terms of air rising over the barrier and being forced to lose some of its moisture through condensation and precipitation on the windward side. The air then descends on the lee side and warms rapidly at the dry adiabatic rate (9.8°C km^{-1}) due to its lack of moisture, giving rise to higher temperatures at the foot of the lee slope than at the foot of the windward slope.

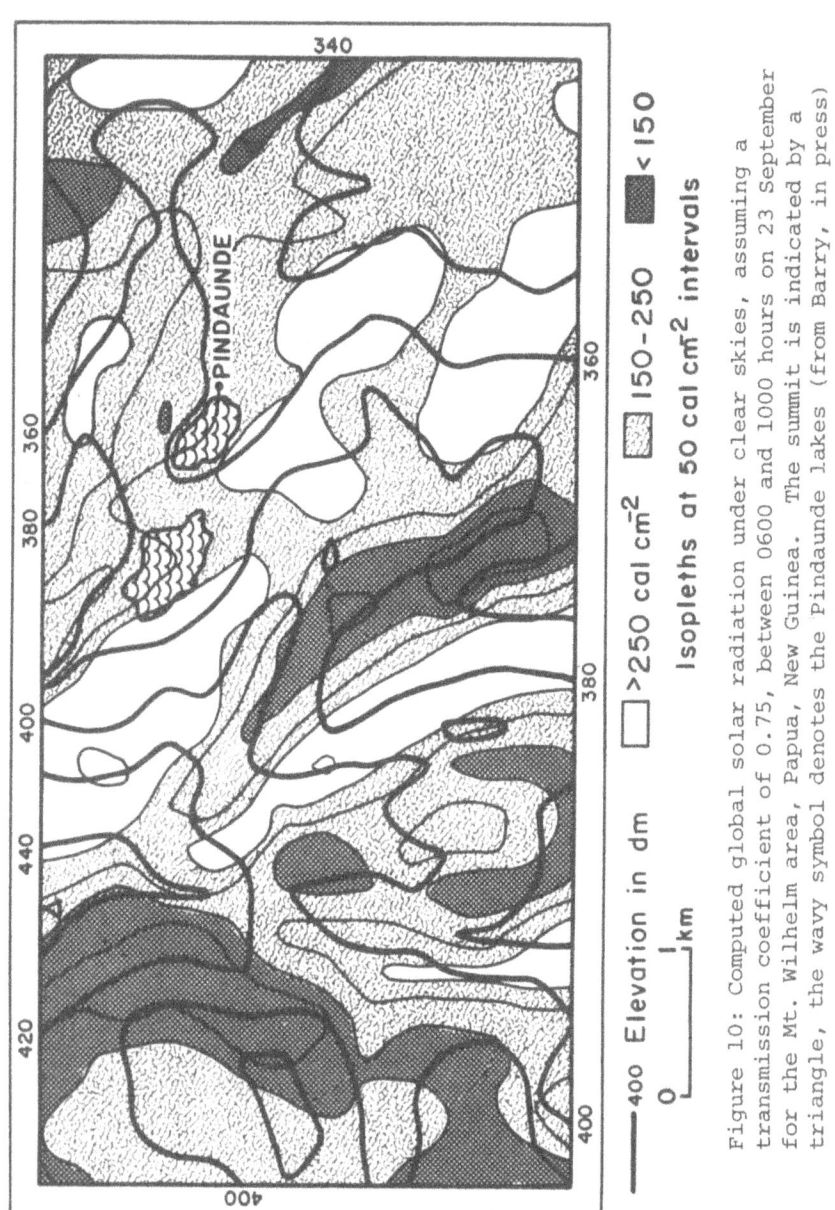

Figure 10: Computed global solar radiation under clear skies, assuming a transmission coefficient of 0.75, between 0600 and 1000 hours on 23 September for the Mt. Wilhelm area, Papua, New Guinea. The summit is indicated by a triangle, the wavy symbol denotes the Pindaunde lakes (from Barry, in press)

Recently it has become apparent that this pattern occurs in rather few cases, at least in the Colorado Rockies (27). More typically the air received in Boulder and Denver has been drawn down from the summit level and there is seldom a loss of moisture through precipitation on the windward slope. Moreover, in more than a few instances, the winds are of the cold (bora) type. Some of these winds are well known for their spectacular consequences in terms of storm damage, as a result of their characteristic strong gustiness. Based on studies by National Center for Atmospheric Research (NCAR) aircraft, the flow characteristics which produce the lee waves represent one type of atmospheric motion involving a small wave amplitude and a short wave length. In contrast, the situation during the January 1971 windstorm in Boulder shows a single major wave with large vertical amplitude (27, 28). The wave trough may shift its location so that the strongest winds may begin at the foot of the slope and work westward up the slope, or they may begin on the upper slopes and then move eastward. The prime control on these wind storms is the existence of a stable layer close to the summit level which blocks the air to the west of the barrier and enables air from above the mountains to accelerate down the leeward slope towards the local pressure trough. Only a narrow belt at the foot of the Front Range is so affected (27).

There are also other wind systems engendered within the mountains due to thermal contrasts along the slopes and longitudinally along the valley. These are mountain and valley wind systems; typically during the day there are gentle up-valley winds and also upward motion along the slopes (29, 30). These conditions may be of considerable importance in terms of fire movement in narrow canyons. At night there is typically drainage down-valley and down-slope, and this is enhanced if there are snow fields or ice bodies in the upper section of the valley.

Concluding Remarks

This survey provides only a broad overview of our knowledge of some of the primary climatic elements in mountain areas. While generally adequate data are available on temperature and precipitation for many mountain areas, there is very little information on characteristics such as solar radiation, wind, or moisture balance in the mountains. Methods of mapping topo- and microclimatic aspects of conditions in mountain regions have received little attention until recently (31). In the late 19th century there was considerable research at mountain observatories in the United States and Europe. Except for Mt. Washington this effort has been largely abandoned in North America in contrast to the situation in Europe.   The value of high-level stations as bench mark stations for monitoring air quality, as well as for studies of the potential of mountain areas for wind energy, apart from more basic scientific questions, is again beginning to be recognized. In view of their growing importance as a year around recreational resource, the time seems ripe for a much greater effort in mountain weather and climate studies. Such studies would be especially timely in the context of the Unesco Man and Biosphere Program, Project 6:   study of the impact of human activities on mountain and tundra ecosystems.

References

1. R.G. Barry and J. Clark, Evaluation of an ERTS-1 Data Collection Platform installed in the alpine tundra, Colorado. *J. Appl. Meteorol.*, 14: 622-626 (1975).

2. A.M. Kahan, *Use of the LANDSAT-2 Data Collection System in the Colorado River Basin Weather Modification Program.* Final Report, U.S. Dept. of Interior, Bureau of Reclamation, Denver, 103 pp. (1976).

3. R.G. Barry, A climatological transect along the east slope of the Colorado Front Range. *Arct. Alp. Res.*, 5: 89-110 (1973).

4. L.W. Larson, Shielding precipitation gages from adverse wind effects with snow fences. Water Resources Research Inst. Rep., Water Resources Ser., 25, Univ. Wyoming, Laramie: 161 pp. (1971).

5. E.F. LeDrew, The energy balance of a mid-latitude alpine site during the growing season, 1973. *Arct. Alp. Res.*, 7: 301-314 (1975).

6. F. Sauberer and I. Dirmhirn, Das Strahlungsklima. In: Klimatographie von Osterreich. *Osterreichische Akad. Wiss.* (Vienna), *Denkschrift* 3(1): 13-102 (1958).

7. M.M. Caldwell, Solar Ultraviolet Radiation as an ecologica factor for alpine plants. *Ecol. Monogr.*, 38: 243-268 (1968).

8. D.M. Gates and R. Janke, The energy environment of the alpine tundra. *Oecol. Planta*, 1: 39-62 (1966).

9. R.J. Hnatiuk, J.M.B. Smith and D.N. McVean, *The Climate of Mt. Wilhelm*. Publ. GB/4, Dept. of Biogeography and Geomorphology Res. School Pac. Studies, Austral. Nat. Univ., Canberra: 76 pp. (1976).

10. G.S. Hope, J.A. Peterson, I. Allison and U. Radok (Eds.), *The Equatorial Glaciers of New Guinea*, A.A. Balkema, Rotterdam: 244 pp. (1976).

11. F. Lauscher, Die Tagesschwankung der Lufttemperatur auf Höhenstation in allen Erdteilen. *60-62 Jahresbericht des Sonnblick-Vereines für die Jahre 1962-64*: 3-17 (1966).

12. H.B. de Saussure, *Voyages dans les Alpes, précédés d'un essai sur l'histoire naturelle des environs de Grenoble.* L. Fauche-Borel, Neuchatel, Tome VII: 393-420 (1796).

13. F. Lauscher, Weltweite Typen der Höhenabhängigkeit des Niederschlags. *Wetter u. Leben*, 28: 80-90 (1976).

14. W. Weischet, Der tropisch-konvektive und der aussertropisch advektive Typ der vertikalen Niederschlagsverteilung. *Erdkunde*, 19: 6-14 (1965).

15. D.E. Colton, Numerical simulation of the orographically-induced precipitation distribution for use in hydrologic analysis. *J. Appl. Meteorol.*, 15: 1241-1251 (1976).

16. J.F. Miller, Physiographically adjusted precipitation - frequency maps. In: *Distribution of Precipitation of Mountain Areas.* Vol. 2, WMO No. 326, World Meteorological Organization, Geneva: 264-277 (1972).

17. J.O. Rhea and L.O. Grant, Topographic influences on snowfall patterns in mountainous terrain. In: *Advanced Concepts and Techniques in the Study of Snow and Ice Resources.* Nat. Acad. Sci., Washington, D.C.: 182-192 (1974).

18. L.M. Hjermstad, The influence of meteorological parameters on the distribution of precipitation across central Colorado mountains. *Atmos. Sci. Pap.* 163, Colorado State Univ., Fort Collins: 78 pp. (1970).

19. G. Østrem and T. Ziegler, *Atlas Over Breer i Sor-Norge* (Atlas of Glaciers in southern Norway). Medd. 20, Hydrologisk Avdeling, Norges Vassdrags-og Elektrisitetsvesen, Oslo: 207 pp. (1969).

20. A. Baumgartner and E. Reichel, Probleme der Alpenhydrologie. In: *14th Internat. Tagung für Alpine Meteorologie*. Rauris, Austria, 1976 (in press).

21. R.G. Barry, Diurnal effects on topoclimate on an equatorial mountain. In: *14th Internat. Tagung für Alpine Meteorologie*, Rauris, Austria, 1976 (in press).

22. L.D. Williams, R.G. Barry and J.T. Andrews, Application of computed global radiation for areas of high relief. *J. Appl. Meteorol.*, 11: 526-533 (1972).

23. P. Wardle, Alpine timberlines. In: *Arctic and Alpine Environments* (J.D. Ives and R.G. Barry, Eds.). Methuen, London: 371-402 (1974).

24. D. Alford, Cirque glaciers of the Colorado Front Range: Aspects of a glacier environment. Unpub. Ph.D. dissertation, Univ. of Colorado, Boulder, 157 pp. (1972).

25. D.T. Lloyd, The Isabelle glacier, Front Range, Colorado, during the 1968-1969 budget years. Unpub. M.A. dissertation, Univ. of Colorado, Boulder, 133 pp. (1970)

26. W.A.R. Brinkmann, What is a Foehn? *Weather* 26: 230-239 (1971).

27. W.A.R. Brinkmann, Strong downslope winds at Boulder, Colorado. *Mon. Wea. Rev.*, 102: 592-602 (1974).

28. J.B. Klemp and D.K. Lilly, The dynamics of wave-induced downslope winds. *J. Atmos. Sci.*, 32: 320-339 (1975).

29. H. Flohn, Local wind systems. In: *World Survey of Climatology*, Vol. II, *General Climatology*, Elsevier, Amsterdam: 139-171 (1969).

30. R. Geiger, Topoclimates. In: *World Survey of Climatology*, Vol. II, *General Climatology*, Elsevier, Amsterdam: 105-117 (1969).

31. M. Hess, T. Niedzwiedz and B. Obrebska-Starkel, The methods of constructing climatic maps of various scales for mountainous and upland territories, exemplified by the maps prepared for Southern Poland. *Geogr. Polonica*, 31: 163-187 (1975).

5

Ice and Snow
at High Altitudes

Malcolm Mellor

Introduction

At sufficiently high altitudes, ice and snow can occur anywhere on the globe, and perennial deposits can be sustained even in tropical latitudes. Where they occur, snow and ice form part of the surficial geology, and exert a major influence on the general environment. Surface energy balance and ground temperatures are affected, ambient light is modified, water is stored in the solid state, and powerful geomorphic processes come into play. These things have profound effects on the landscape, on physical processes in soils and rocks, on hydrology, on plant and animal life, and on human activities.

The high altitude environment is basically a ground surface at some relatively high level in the atmosphere, and this implies reduction of air pressure and optical air mass. Within the troposphere, mean air temperature decreases with altitude at about $6.5^\circ C$ km^{-1}; atmospheric pressure and optical air mass drop to about half the sea level values at an altitude of 5.2 km or so. Temperature is obviously important in determining minimum altitudes for the existence of snow and ice; it is also significant that snow and ice in nature always exist at high homologous temperatures, so that their properties are very sensitive to temperature. The decrease of air density and optical air mass means that direct solar radiation suffers less attenuation, and the intensity and spectral distribution of incident radiation are not the same as at sea level. Decreased atmospheric pressure affects the partial pressure of water vapor, which in turn influences evaporation, vapor diffusion, and sintering. However, this is probably not very important, since for practical purposes saturation vapor pressure for a given surface curvature is controlled by temperature. There might also be more subtle altitude factors, for example variations of potential gradient in the atmospheric electrical field, but this is only speculation.

In addition to the intrinsic altitude factors, an elevated land mass can influence conditions by virtue of its form and dimensions. Movements of air masses are affected, with consequent influence on precipitable moisture, cloud distribution, wind speed, and wind direction. In true mountain areas, as distinct from high plateaus, the topography provides slopes with varying exposure to sun and wind. Slopes also provide conditions for flowing and sliding of snow and ice.

However, while the high altitude environment has many special characteristics, there is not much evidence to indicate that snow and ice are themselves very different at high altitudes than at low altitudes. Certain surface features, such as *nieves penitentes*, are associated almost exclusively with high altitude snowfields and glaciers, but these seem to be features developed by intense solar radiation at high altitudes and low latitudes, especially in central Asia and South America. In any event, since there have not been many scientific studies dealing explicitly with snow and ice at high altitudes, we have to draw on general knowledge gleaned in the laboratory and in a variety of natural settings. The latter include the extensive high altitude snowfields of Antarctica and Greenland, but in more spectacular mountain regions science competes poorly with the seductive delights of climbing and skiing.

This short review deals with some basic properties of snow and ice that govern environmental processes and research procedures.

Ice and Snow

Ice is the solid form of water. Snow is either a suspension or a compact of ice particles, with air, and sometimes water, in the interstices. The ice that occurs in nature is termed ice Ih, but there are other kinds of water ice. Deposition of water vapor on very cold substrates can produce cubic ice (ice Ic) or vitreous ice, while various combinations of high pressures and low temperatures can produce eight high pressure polymorphs.

An ideal crystal of *ice Ih* has the oxygen atoms of the water molecules arranged tetrahedrally at a separation of 2.76 A, which gives rise to hexagonal symmetry in the crystal There are two hydrogen atoms adjacent to each oxygen atom, but only one hydrogen atom lies between each pair of oxygen atoms, and each molecule is oriented so as to direct its hydrogen atoms towards two of the four neighboring oxygen atoms. Molecules lie close to a set of parallel planes, which are the basal planes of the crystal. A unit cell in the crystal structure contains four molecules, and its dimensions (about 4.5x4.5x7.4A) correspond to a density of 0.917 Mg m^{-3} at 0°C.

(Incidentally, low density is one of the most important characteristics of ice; if ice were denser than water, this would be a very different planet.)

Ice occurs in a variety of natural forms. In the *atmosphere*, ice is formed in clouds by vapor deposition or freezing of supercooled droplets, with heterogeneous nucleation around ice nuclei in both cases. Under appropriate conditions, embryonic ice particles grow by vapor deposition, by particle aggregation, or by accretion of supercooled water, eventually reaching precipitable sizes. Crystal habit is controlled largely by temperature during the formation and growth stages, and riming can result from collisions with supercooled water droplets. *Snow crystals* reaching the ground range from simple needles and prisms to intricate planar and spatial dendrites. Riming may occur to various degrees, with *hail* representing the most extreme form. Solid precipitation can be deposited directly after falling vertically, or it can be put down as a wind-blown deposit after horizontal transport by turbulent diffusion and saltation. Ice can also form directly from the atmosphere on ground surfaces, plants, and structures. Vapor deposition on solid surfaces gives *frost* or *hoar*, freezing of small supercooled droplets gives bubbly *rime*, and freezing of large supercooled droplets or rain on cold surfaces gives clear *glaze*.

Most of the seasonal or perennial ice encountered at high altitude is *deposited snow*, or *glacier ice* formed from snow. After deposition (at densities ranging from less than 0.1 to as much as 0.4 Mg m^{-3}), snow is buried and compacted by succeeding snowfalls, and eventually it can turn to ice. This can be either a completely dry process, or one that involves melting, water percolation, and refreezing. Overburden pressure increases with depth at rates of 0.02 to 0.07 bar m^{-1}. Pores in dry snow seal off to form closed bubbles at a density of about 0.8 Mg m^{-3}, and this is usually taken as the snow/ice transition. The size and state of a glacier are determined by its energy balance and its mass balance. The mass balance for a complete glacier involves *input* by direct precipitation, condensation, wind transport, and avalanches, versus *loss* by melting and runoff, evaporation, wind transport, and ice flow (feeding avalanches or calving sites in lakes).

The mass balance for any columnar element of the glacier system is determined by the flux divergence equation for the element.

Ice also forms by *freezing of water* in lakes, streams, soils, rocks, and groundwater seepages. Freezing at the surface of a still water body is an orderly process. Because water has its maximum density at +4°C, water stratification is stable; nucleation and crystal growth can, therefore, initiate near the air/water interface with warmer water below. Ice growth proceeds at a rate controlled largely by heat conduction through the ice cover. Growth and decay can be related to "degree-day" summations, with growth rate decreasing as ice thickness increases. Air bubbles are trapped within the ice, soluble impurities concentrate at crystal boundaries, and slow growth produces preferred crystal orientation. The freezing process is similar in quietly flowing water, with some delay occasioned by vertical mixing. In swiftly flowing water, crystals nucleated near the surface can be distributed through the stream by turbulent diffusion and the water can be supercooled. The crystals sometimes accumulate on obstructions to form a spongy mass known as *frazil*. Boisterous streams also create spray icing on rocks and plants, and develop curtains of icicles at waterfalls. When water issues from the ground or from snowpacks at very low rates it freezes in thin layers, building up icings (*aufeis, naled*) on level or sloping ground, and icicles where the water drips vertically.

Freezing in soils and rocks is a complex process. In coarse-grained or coarsely porous materials, "bulk" water in the large pores freezes at close to 0°C, with some freezing point depression when there are soluble impurities. However, in material with fine grains or fine pores, some water remains unfrozen at temperatures below 0°C, and for a given soil there is a characteristic relationship between unfrozen water content and temperature. Below -5°C, the unfrozen water content at a given temperature correlates with the specific surface area of the internal structure of the material, which can be measured by liquid or gas adsorption. Clay soils, and rocks formed from clay, have the greatest capability for remaining unfrozen at low temperatures. Seasonal freezing of pore water in soils and rocks is the basis of frost heave, frost weathering, and formation of patterned ground. Ice in frozen ground is found within the pores, in discrete partings or lenses formed by water migration and in wedges formed by percolation into contraction cracks. In permafrost there may also be bodies of buried ice that are remnants of glacier ice, snow beds, lake ice, and such like. A special kind of permafrost found in high mountains is the rock glacier, a slowly creeping scree that has interstitial or underlying ice.

Formation of ice in cells and tissues of animals and plants is perhaps even more complicated, at least in its effects. As with soils, there are problems of interfaces and freezing in small "pores," but the chemistry is usually more complicated. Cooling rate is an important factor in determining whether ice forms within as well as outside the cells, and it controls crystallization rate, salt concentration, and water movements. An important practical consideration is the extent to which living matter is damaged, mechanically or chemically, by freezing.

Impurities in Snow and Ice

The snow and ice on high ice caps and high mountains must be about the purest water that occurs anywhere in large quantities, but there are some impurities, even in the most remote and apparently unpolluted regions. Here we are considering only "virgin" snow and ice, and not ice formed on stagnant ponds, or ground ice.

For a start, snow contains small amounts of the stable isotopes oxygen 18 and deuterium. The ratio of $H_2^{18}O$ to the common $H_2^{16}O$ decreases slightly with the temperature of snow formation, providing an indicator of seasonal and secular temperature variations for the formation site, and offering a means of dating snow strata.

Snow also contains radioactive isotopes, both from natural sources (nucleogenesis, cosmic ray activity) and from human activity (weapons tests, power plants, nuclear industries); these can be used for dating "snow ice" over long periods. Isotopes may be in the ice molecules, in adsorbed or included air, in dissolved impurities, or in insoluble particles. Tritium (^3H) occurs naturally, but there have been great additions from nuclear devices and industries since 1954.

Chemical contamination occurs even in the absence of local human activity. Many inorganic elements and ions have been found in "clean" snow, with sources that include terrestrial dust and ocean spray, volcanic gases and dusts, plant emanations, combustion products from fossil fuels, fumes and dust from industrial processes, and dust and spray from road salts. Organic contaminants include hydrocarbons from combustion, organic synthesis, industrial processes, plants, volcanoes, and such like, as well as identifiable compounds such as DDT (dichloro/diphenyl/trichloroethane) and PCBs (polychlorinated biphenyls).

Insoluble particles can be introduced both by direct deposition and by incorporation into precipitation during condensation, nucleation, or crystal growth. Sources include windborne dust from terrestrial soils and rocks, ash from volcanoes and industrial processes, plant pollens, plant fragments, phytoliths, and extraterrestrial particles.

Certain algae, fungi, bacteria, molds, and insects can live on snow surfaces where light is plentiful and temperatures are near the melting point. Organic tinting can give shades of red, green, yellow, or blue.

Ice formed from snow compaction contains air bubbles. These provide samples of atmospheric gases, representative of the time and place of closure for the bubbles. In addition to the basic oxygen and nitrogen, CO_2, CO, CH_3, and Ar have been analyzed.

Very clean snow is usually slightly acidic, with pH between 4 and 7. At high altitudes in Greenland and Antarctica, the pH of meltwater is typically between 5 and 6, which is a bit below the desirable minimum for drinking water.

Movement and Deformation

Snow and ice deform quite easily at low stress levels, and gravity body forces are enough to keep both snow and ice continually on the move. The only exception is when dense ice is in a purely hydrostatic state of stress, a condition that doesn't occur much in nature, except in frozen ponds and lakes.

Like any material, ice deforms elastically, with a true modulus of about 90 kilobars that is not very sensitive to temperature. It also creeps when subjected to deviatoric stress, and there is no known lower limit of stress below which it completely stops creeping. However, the relationship between strain rate and stress is nonlinear, and for shear stresses below about 1 bar (give or take a factor of 2 to account for temperature variations) the creep is almost imperceptibly slow. Above the 1 bar stress level, strain rate increases approximately as the cube of stress, so that there can be an appearance of plastic yield. As a matter of interest, deviatoric strain rate at the 1 bar stress level is between 10^{-9} and 10^{-8} s^{-1} with common temperatures (say -2 to -10°C).

This means that in nature, snow and ice creep very rapidly until the deviatoric stress drops to a certain level; thereafter creep becomes much slower. For example, under normal conditions a glacier adjusts its thickness and slope so that the maximum shear stress, which generally occurs at the bed, is somewhere near the 1-bar level. If something happens to increase thickness or surface slope, the ice deformation speeds up so as to restore equilibrium, and vice versa. As another example, freshly fallen snow compacts very rapidly under its own weight until its density has increased enough to relieve high stresses within and between the constituent ice crystals. The initial volumetric strain rates of fluffy new snow are of the order of 3×10^{-6} s^{-1}, and they can even exceed 10^{-4} s^{-1}. By contrast, well settled dry snow continues to compact at rates only of the order of 10^{-8} s^{-1} or less.

The significance of this behavior is that some of the important characteristics of snowpacks and glaciers are reasonably predictable. In flat-lying dry snow deposits, the relationship between density and depth, and between overburden pressure and depth, have characteristic forms, which only vary within fairly narrow limits for different regions, mainly according to temperature conditions. On a snow slope, the downslope creep deformation is more or less predictable. To a first approximation, the shear strain rate of well settled snow is invariant with depth (so that plane cross sections remain plane), and under typical mountain conditions the shear strain rate divided by the tangent of slope angle is of the order of 3×10^{-8} s^{-1}. In the case of a glacier that is leading an orderly life, some aspects of its geometry are predictable, to the extent that the product of ice depth and local surface slope is fairly constant--ice depth x unit weight x slope is of the order of 1 bar.

Temperature has a strong effect on the relationships between strain rate and stress, creep rates decreasing as temperature decreases. When ice is deforming at temperatures below -10°C, the variation of strain rate with temperature can be described by an Arrhenius relation that has a creep activation energy of 60 to 70 kJ mole^{-1}. However, above -10°C creep increases sharply with temperature; at 0°C creep is at least an order of magnitude faster than at -10°C.

So far we have been considering internal deformations, but both snow and ice deposits can slide over the supporting ground surface when the interface temperature is close to the melting point. This process is not so well understood; in fact, the general behavior of wet snow and wet ice is rather complicated and in need of further study. The consequences of sliding are certainly significant. The sliding, or "gliding," of snow on a slope can deflect, flatten, pluck, or completely uproot vegetation, an effect that can be seen in springtime from ski lifts in the Alps. The sliding of glaciers, which involves large-scale surface roughness, is a major process for erosion and forming of the landscape. It also seems to be the mechanism for major glacier surges.

Ice that contains dispersed soil particles usually deforms more easily, or creeps faster, than clear ice. Increasing concentration of solid particles first produces a corresponding increase of creep rate, but when the particles begin to dominate the mixture, creep is gradually suppressed, especially in granular soils where internal friction can be mobilized.

Creep Rupture and Brittle Fracture

Deformation sometimes culminates with the snow or ice breaking. We make a distinction between brittle fracture, which is preceded by deformation that is largely elastic and reversible, and creep rupture, which is a separation resulting from progressive loss of deformation resistance through dissipative processes. This loss of resistance is illustrated by peaking in a stress/strain curve from a constant rate test, or by acceleration (or tertiary creep) in a constant stress creep test.

In snow and ice, true brittle fracture has to involve very rapid loading, of the kind that might be produced by jumping on it, hitting it with an ice-axe, blasting it with an explosive, or impacting it with a projectile. The spontaneous ruptures like crevasse formation, thermal cracking, snow quakes, or avalanche release all follow slow loading and creep. The fact that the material breaks with a snap, crackle, or pop does not necessarily mean that it is brittle fracture--that is just the release of elastic strain energy.

The strength characteristics of snow and ice are complicated. Consider first the behavior of solid ice, which can be assumed incompressible. Under any constant deviatoric stress, it will eventually fail and we can think in terms of "time to failure" or "strain to failure," noting that for practical purposes the time to failure can be virtually infinite if the stress is low enough. Under any constant strain rate, there is a maximum stress that can be sustained, and this we can call the strength, whether the material actually breaks or not. Both of these cases are covered by the statement that there is a maximum limiting value for the ratio of stress to strain rate at any given temperature.

The limiting stress/strain-rate relation of ice is easiest to discuss for uniaxial stress states, either tension or compression. If we plot on logarithmic scales the maximum stress against strain rate, or alternatively the minimum strain rate against stress, the trend at very low values of stress and strain rate is toward viscous behavior, i.e., strain rate proportional to stress. At higher values, strain rate is proportional to some power of stress, as mentioned already in the discussion of creep deformation. The exponent of this power relation tends to increase with stress and with strain rate, towards an infinite value as the limiting strength is approached at high strain rates. In tension, this limit is apparently reached around 10^{-7} s^{-1} at a stress of about 20 bars. For compression, there is now some evidence to indicate that the limit has not quite been reached at 10^{-1} s^{-1} and 150 bars for a temperature of $-7^{\circ}C$. Incidentally, compressive strength varies with temperature, while high rate tensile strength is almost independent of temperature.

As far as we can tell, porous ice and snow behave in a qualitatively similar fashion as long as we consider the relationships between deviatoric components of stress and strain rate for constant volume deformation. However, we cannot expect equivalent behavior in snow if bulk stress and volumetric strain are changing the density of the material as the deviatoric straining progresses.

This brings up the next complication. It is not sufficient to have a relationship between the deviatoric components of stress and strain rate. We need to know how strength is affected by all the stress components in a multiaxial stress field. The required relationship, known as a failure criterion, is a critical relation of principal stresses that defines the limits for failure.

In solid ice, the yield stress for creep rupture is virtually unaffected by hydrostatic pressure, but the fracture strength for high loading rates is significantly dependent on bulk stress. The failure envelope changes with strain rate, as can be seen from variation in the ratio of uniaxial compressive strength to uniaxial tensile strength. In snow, it is even more difficult to express effects of bulk stress in formal terms, because compression produces time-dependent density changes.

In frozen soils that have high ice content, such that grains are separated by ice, strength characteristics seem to be qualitatively similar to those of clear ice. In granular soils that have low ice content, or high dry density, the frictional component of shearing resistance ought to provide a finite yield strength. In tension, all ice-bonded soils are stronger than in the unfrozen state for any rate of loading, but creep can occur at any stress.

Compressibility of Snow and Ice

A great deal of the world's ice, especially that originating at high altitudes, is formed by the dry compression of snow, and compressibility exerts a controlling influence in many practical problems.

In talking about compressibility we refer to the relationship between hydrostatic stress and corresponding volumetric strain or, in a general multiaxial stress state, between the isotropic components of the stress and strain tensors. This can involve elastic compression, as characterized by the bulk modulus, or nonlinear viscous compression brought about by creep, or abrupt compression involving structural collapse, or phase transformation.

Elastic compressibility, which is not significantly dependent on rate or temperature, is easy enough to describe. The bulk modulus of coherent dry snow increases with density from about 400 bars at a density of 0.3 Mg m^{-3} up to roughly 20 kilobars at 0.6 Mg m^{-3}. The increase then continues at a lower rate, up to a value of about 90 kilobars for solid ice.

The viscous compressibility is less easy to define. Relationships between hydrostatic pressure and specific volume, which are sometimes known as "equations of state," are heavily rate-dependent and temperature-dependent. When pressure is plotted against specific volume (which is actually a dimensionless measure of density), there are innumerable compressibility curves with rate and temperature as parameters. A high rate upper limit is set by the relation for shock compression—the so-called shock adiabat or Rankine-Hugoniot equation of state. If ice and snow had a finite yield stress, a lower limit would be set by curves for very slow isothermal compression at high temperature. In principle, we do not have such a limiting isotherm for snow and ice, but for most practical purposes there actually is a limit zone that represents slow compaction under gravity body forces. This gives the maximum density that is reached after many months of exposure to a given stress level.

In some respects it makes life easier to regard irreversible viscous compression of snow as a quasi-plastic process in which snow collapses to a higher density in response to addition of pressure. This is not too unrealistic for certain problems, since addition of pressure produces a volumetric strain rate that decays with time towards an asymptotic limit. If the "half-life" for this pseudo-equilibration is short compared to the interval between stress changes, a quasi-plastic compression modulus can be defined in terms of the stress/density gradient. For dry densification on snowfields and glaciers, such a modulus might range from the order of 10^{-4} bar at a density of 0.1 Mg m^{-3}, to about 0.7 bar at a density of 0.6 Mg m^{-3}, where the snow grains are near the "close packing" limit. There is a further modest increase in the modulus, up to about 2 bar at a density of 0.8 Mg m^{-3}, where snow makes the transition to impermeable ice.

When snow is transformed to ice by dry compaction, the pore structure seals off to form discrete air bubbles at a density of about 0.8 Mg m^{-3}. Further densification requires that the air in the bubbles be compressed in accordance with the gas laws, with the ice itself deforming in order to close the bubbles.

High speed compaction of snow is involved in a variety of engineering problems, including avalanche impact. These problems typically are approached by consideration of mass and momentum conservation across a plastic wave front, the pressure rise being determined by the high rate compressibility characteristic, or Hugoniot. Shock adiabats for ice are not of much significance in the present context unless we are worried about explosions or projectile impact. However, it might be noted that simple impact stress for ice reaches 1 kilobar at an impact velocity that is only about 30 m s^{-1}. The Bridgman characteristics for extreme isothermal compression are not very relevant either, except insofar as depression of the melting temperature is concerned (about 0.0074°C bar^{-1}).

Saturated frozen soils compress more or less elastically up to the very high stress levels where phase transition occurs in the pore ice. However, pressure increases the thickness of unfrozen water films around soil grains in accordance with the Clausius-Clapeyron equation, and this is thought to affect frost heaving and particle migration. Incidentally, frost heaving can create very high pressures; pressures of the order of 100 bars have been measured in low permeability soils, and higher values are theoretically possible. Ice segregation during freezing of clays can also create pore water tensions of comparable magnitude, producing overconsolidation without external reaction.

Elastic Waves

Elastic wave propagation is the basis of seismic methods for subsurface exploration and depth sounding. Wave velocities are calculable from the elastic moduli, and in principle wave attenuation for a given frequency should be calculable from values of the loss tangent measured in laboratory studies of anelastic behavior. However, for practical purposes it is safer to rely on direct measurements.

The velocity of a dilatational wave in ice is about 3.8 km s^{-1}, which is more than 2.5 times the velocity in water, and well over 10 times the velocity in air. It decreases as porosity increases, to about 3.4 km s^{-1} at the snow/ice transition. It is around 2.5 km s^{-1} at a snow density of 0.6 Mg m^{-3}, and it has dropped to less than 1 km s^{-1} at a density of 0.3 Mg m^{-3}. There is not much variation with temperature.

The shear wave velocity for ice approaches 2 km s^{-1}, and it decreases steadily to about 1.6 km s^{-1} at a density of 0.65 Mg m^{-3}. Below this "close-packing" density it decreases more steeply, to about 0.4 km s^{-1} at a density of 0.35 Mg m^{-3}.

In both soils and rocks, there is a significant increase in wave velocity as pore water freezes. In coarse-grained soils or high porosity rocks, the change is abrupt, taking place close to 0°C if there is not much soluble impurity. In fine-grained materials, especially clays, the change in wave velocity spreads over a range of temperature, since the water freezes progressively as temperature decreases. In a saturated clay, the seismic velocity can be three times higher in very cold soil than in unfrozen material.

Surface Friction

Low surface friction is one of the most agreeable properties of snow and ice. In discussing kinetic friction of snow and ice, it is as well to forget the classical Amontons/Coulomb laws of dry friction, because lubricating water films, or liquidlike surface layers, play an important role, and the friction coefficient turns out to be a function of bearing pressure.

The old ideas about pressure melting are not now thought to be very relevant unless temperature is close to 0°C. It is true that high pressures can be developed by some sliders, such as skate blade edges, but at -10°C the phase change from ice Ih to water does not take place until the pressure exceeds 1 kilobar. Present ideas center around frictional heating, leading either to actual melting or to some thermally activated orientation process that amounts to much the same thing.

For any slider, the power density, or energy generation rate per unit area, is given by the product of frictional stress and sliding velocity. This, together with the thermal properties of the slider and ice material, determines the rate of temperature rise and the rate of melting at the interface. The total amount of water produced, which relates to the thickness and continuity of the lubricating film, depends on the power density, the thermal properties, and the duration of contact. This is a simplified picture, but it serves to explain most of the experimental data, which tend to be a bit confusing at first sight. The general situation is that friction tends to decrease with increasing velocity, increasing temperature, increasing pressure, and increasing slider length. There are, of course, numerous complications connected with surface roughness, soft slider surfaces, hydrophilic sliders, excess water, and domination by one variable. There can also be experimental confusion, for example when snow compaction resistance is lumped with true friction effects. The slider materials that most consistently provide low friction are mainly hydrophobic polymers that develop only weak molecular forces at the interface, as indicated by high contact angles with water. Some of these apparently have the ability to orient and shed their surface molecules, providing low friction even at very low temperatures, where a lubricating water film is hard to develop.

Static friction on snow and ice tends to be confused with adhesion, and it might be well to regard it as the isothermal limit of kinetic friction when sliding velocity tends to zero. If a slider stands still for some length of time, it can develop interfacial bonds that produce a tensile connection whose strength is proportional to area. True static friction on cold ice and snow ought to approximate to kinetic values for unlubricated conditions, as developed at very low speeds and very low temperatures. At 0°C, there should not be much difference between static and kinetic friction if surface tension effects in the interstitial water are small.

Adhesion is important, especially in relation to icing on plants, trees, structures, power lines, and so forth.

When initial contact is made under "dry" conditions, the strength of an adhesive bond can be expected to increase exponentially with time up to an asymptotic limit, with a rate constant controlled by temperature, temperature gradient, contact pressure, and surface properties for both materials. Bond strength can be expected to increase with decreasing temperature, and it will vary with the surface properties of the host material, strongly hydrophobic materials being least likely to provide strong bonds. Upper limits to bond strength in shear and tension are set by the strength of the ice material or the host material, whichever is weakest.

Thermal Properties

Of all the thermal properties of snow and ice, perhaps the most important is the *melting temperature*, which we usually take as $0^{\circ}C$ for pure bulk ice. A glance at the pressure/temperature phase diagram shows that we have plenty of choice if we do not like that value, but a more important practical matter is that pure water in tiny droplets, adsorbed layers, fine capillaries, and other fine pores can remain liquid to very low temperatures. In fine-grained soils and rocks, especially those containing clay, an appreciable fraction of the water can remain unfrozen at temperatures down to $-10^{\circ}C$ or so; this permits water movement under any potential gradient, with immediate freezing when the water reaches some place where it can take on "bulk" properties. This is a basis for frost heaving and frost damage to rocks, and one imagines that it must be a factor in the frost hardiness of plants.

Specific heat is not particularly interesting, but we do have to know it. Under the old system of units it was usually enough to remember the number 0.5, for either English or metric units. Now we have to think of the specific heat for pure ice at constant pressure and $0^{\circ}C$ as 2.12 J g^{-1} K^{-1}. Specific heat at constant volume is about 3% less at $0^{\circ}C$. Over the range of ordinary environmental temperatures, C_p decreases with temperature at about 0.34% per degree. With impurities present, the *apparent* specific heat is different from the pure ice values.

The *latent heat* of ice provides tremendous storage capacity for energy, something which we are becoming aware of in the experimental design of heating and air conditioning systems for buildings. The same thing applies in nature--it takes a lot of energy to banish snowfields and glaciers. At $0^{\circ}C$ and standard atmospheric pressure, the latent heat of fusion for pure ice is 333.5 J g^{-1}. It decreases with temp-erature roughly 1% per degree, but that is not of much significance here. The latent heat of sublimation at 0°C and standard pressure is 2838 J g^{-1}, and it is not much affected by temperature (latent heat of vaporization increases with decreasing temperature).

The *thermal conductivity* of snow and ice is significant in many ways in the high altitude environment. In simple terms, we can think of snow and ice as insulating layers that block heat conduction from above and below. Surface temperature waves, such as diurnal or annual cycles, suffer phase lag and attenuation, and geothermal heat has to pass through superimposed ice and snow, setting up its own gradient. Ordinary seasonal snow cover is often sufficient to keep the surface ground temperature close to 0°C, providing snug winter conditions for plants and animals. A floating ice cover on ponds or lakes blocks heat loss from the water below, and growth rate decelerates as the ice thickens. Thick perennial deposits of dry snow and ice soon filter out higher frequency temperature waves, and even the annual temperature cycle is completely damped out below a depth of 10 to 15 m. This means that conduction is incapable of letting the bed of a thick deposit know whether it is winter or summer. It also gives the deep layers a capability for "remembering" mean annual surface temperatures, so that in flat-lying or very slowly moving ice, climatic changes could be reflected in deep temperature profiles. In a dry snow accumulation area where there is appreciable downslope movement, this "memory" of mean surface temperature can lead to a negative temperature gradient in the deep layers, i.e., the ice gets colder as depth increases because of downhill travel and burial.

The thermal conductivity of solid ice is fairly straight-forward. At temperatures between 0 and -10°C it is about 2 J m^{-1} s^{-1} K^{-1} and there is a slight increase as temperature decreases within the ordinary environmental range (about half a percent per degree).

In snow the situation is very different, for heat transfer involves solid conduction, conduction in air spaces, vapor diffusion and convection in the voids, and radiation between grains. We therefore use an *effective* thermal conductivity to cover these complications, and it can vary greatly with the structure of the snow and the method of measurement. In low density snow with density about 0.1 Mg m^{-3}, the effective thermal conductivity is of the order of 10^{-1} J m^{-1} s^{-1} K^{-1}. It might average 0.2 J m^{-1} s^{-1} K^{-1} at a density of 0.25 Mg m^{-3}, and for higher densities it increases roughly as the square of snow density. Although conduction in the air spaces is not very important, and the conductivity of ice is not strongly temperature-sensitive, we still find that the effective thermal conductivity of snow decreases appreciably with decreasing temperature. This is probably because of the role of vapor diffusion.

Vapor diffusion in snow is an important process, both as a means of transferring heat and as a mechanism for transferring mass. Saturation vapor pressure increases with temperature, and a temperature gradient in snow sets up a gradient of vapor pressure, so that molecules are transferred from grain to grain, carrying along latent heat. Between 0 and -10°C, the diffusion coefficient of water vapor in free air at standard pressure is about 20 mm^2 s^{-1}, but the diffusion coefficient for vapor transfer in typical snow is a good deal higher, in the range 70 to 100 mm^2 s^{-1}. This means that the effective thermal conductivity attributable solely to vapor diffusion is about 8×10 J m^{-1} s^{-1} K^{-1}. Lower values can be expected in colder snow (65 mm^2 s^{-1} has been measured between -7 and -17°C), and values should be higher at high altitudes. In snow of density 0.25 Mg m^{-3}, vapor diffusion could account for about half of the total heat transfer, and in low density snow, say 0.1 Mg m^{-3}, it is likely to dominate the heat transfer process.

Forced convection can greatly intensify heat transfer and vapor transport in snow. We do not normally think of this as applying to natural snow covers, but air pressure gradients could be set up in snowdrifts when strong winds are blowing.

The only other thermal property we need mention is thermal strain, or *expansion coefficient*. When pure water freezes at 0°C, it undergoes a volume expansion of about 8%; this allows it to float, to burst water pipes and radiators, and possibly to damage rocks. In the solid state, ice has a coefficient of linear expansion of approximately 5×10^{-5} C^{-1} at 0°C, which is about an order of magnitude higher than typical expansion coefficients for low porosity rocks. The expansion coefficient of ice decreases with temperature at a rate of 0.4% per degree.

Optical Properties

Bubble-free ice is transparent to visible light to about the same extent as pure water. Entrant light is absorbed and refracted ($n = 1.31$), and a smooth surface has specular reflection characteristics. When ice has bubbles or internal cracks, it scatters and reflects light and, like other dielectrics, it turns white when finely powdered, becoming very effective in blocking transmission of radiation.

The main things to be considered here are the transmission
and attenuation characteristics, which control the penetration
of light, and the related reflection properties, which
govern the radiation balance, and thereby influence the
local climate.

There is some uncertainty about the exact details of
internal scattering and attenuation in snow, but for present
purposes we can assume that diffuse incident radiation
attenuates exponentially with distance according to the
Bouger-Lambert law. The exponential attenuation constant,
or extinction coefficient, of this law is the sum of the
true absorption coefficient, which dominates in clear ice,
and the scattering coefficient, which is dominant in snow.
For clear ice, the extinction coefficient is from less than
0.1 to more than 1.0 m^{-1}. In bubbly ice, the range is from
about 1 to over 10 m^{-1}. The highest extinction coefficients
occur in dense fine-grained snow, where values can exceed
100 m^{-1}. In snow, the extinction coefficient ought to
increase with density, but because there is little experimental information we have to interpolate between measured
values for dense deposited snow and deducible values for low
density suspensions, as represented by falling and blowing
snow. Taking this approach, we find approximate proportion-
ality between extinction coefficient and density, from
values of the order of 10^{-3} m^{-1} in light snowfalls to values
exceeding 10 m^{-1} in fairly compact deposited snow.

Grain size has a strong effect on attenuation, as can
be appreciated by contrasting the transparency of a layer of
ice cubes with an equally thick layer of powdered ice. Both
layers will have about the same density if they are well
shaken, but the layer of ice cubes will transmit more light
and reflect less. Extinction coefficient varies inversely
with grain size, but the exact nature of the relationship
has not been determined.

The other important matter is the reflectance of snow
and ice, which we call the albedo when the reflectance is
being integrated over a fairly broad band of the spectrum.
We have just mentioned that clear ice has very low extinction
coefficients for visible radiation, and this means that a
thick mass of clear ice with a smooth surface will be a poor
reflector for diffuse or normally incident light. For
normal incidence of visible radiation, the albedo is about
2%. When the ice has bubbles and cracks that produce internal
scattering, the albedo can reach 50% or more. At the other
extreme, a deep layer of very fine dry snow can have an
albedo approaching 100% under diffuse illumination. In old
coarse-grained snow, especially wet snow, albedo can drop

back to the 60 to 80% range. In other words albedo, like
extinction coefficient, is inversely related to grain size.

Albedo is simple enough under diffuse illumination,
since the snow behaves as a Lambert surface, but it is
complicated a bit by directional radiation and by directional
features on the snow surface. Even without directional
features, the albedo of a level surface is expected to
increase slightly as solar altitude decreases. This obviously
has some significance in mountainous terrain.

So far, we have not considered the variation of attenuation and reflection with wavelength. This might not seem
of much consequence in the present context, since it takes
very little snow to paint the landscape white and shut out
light from the ground below. However, it is relevant to the
applications of remote sensing and satellite imagery.

In clear ice, the absorption coefficient increases with
wavelength through the visible spectrum (about 0.04 m^{-1} at
0.4 μm to 0.6 m^{-1} at 0.7 μm), but it might also increase a
bit on going into the ultraviolet. Absorption increases
with wavelength in the near infrared (perhaps of the order
of 10^2 m^{-1} at 1.3 μm). There are a number of absorption
bands in the near and far infrared. In bubbly ice, attenuation
does not seem to vary much through the visible, and the same
is probably true for snow. The experimental data are
conflicting, some showing an increase of extinction coefficient
with wavelength, and some a decrease.

The reflection of normally incident radiation from
clear ice is very small throughout the visible and infrared
ranges, but there are a few "peaks" related to absorption
bands in the infrared. The spectral reflectance of fresh
dry snow does not vary much through the visible range, but
for wet or coarse snow there seems to be a decrease towards
the red end of the visible. There is an overall decline in
reflectance through the near infrared for all types of snow,
but there are some local peaks of reflectance around 1.1,
1.8, and 2.25 μm, and these vary in height according to the
grain size of the snow; the finer the grains, the higher the
reflectance. Above 3 μm, reflectance is very small.

Finally, the long-wave emmissivity might be mentioned,
since it relates to the surface heat balance. Solid ice is
pretty much a "black body," with an emissivity of about
0.97, and wet snow is probably similar. However, skimpy
experimental data indicate that the long-wave emissivity of
snow decreases with decreasing grain size and with decreasing
temperature.

Electrical Behavior

The electrical behavior of ice and snow may seem rather irrelevant, but it is becoming increasingly important for subsurface exploration and remote sensing.

Ice is a dielectric, and we can discuss the complex dielectric constant in terms of its real and imaginary parts, the relative permittivity and the loss factor, respectively.

The static permittivity of pure ice is about 100 at typical environmental temperatures, and perhaps a bit less at 0°C (it decreases with temperature at about 0.4% per degree). For comparison, liquid water has a static permittivity of about 88 at 0°C. After frequency has increased to a certain value, permittivity drops quite sharply over a two-decade frequency range, to a high frequency value of about 3.2, which does not vary much with temperature. This drop, or Debye dispersion, is centered about a frequency of some 10^4 Hz at 0°C, and approximately 10^2 Hz at -40°C. By contrast, the dispersion frequency for liquid water is about 10^{11} Hz, so that there is a big difference in the permittivities of ice and water over much of the radio and microwave range.

The permittivity of dry snow follows the same kind of trend, but it varies with snow density and with grain structure. The static permittivity increases rapidly with density, from a lower limit of 1.0 for zero density, up to about 100 at the density of ice. The high frequency permittivity increases from the limit value of 1.0 more slowly, eventually reaching the ice value of about 3.2. Low frequency permittivity can increase by a factor of 2 during a period of sintering at constant density. In wet snow, high frequency permittivity increases with liquid water content. Water content has been found to have linear correlation with permittivity, or with the difference of wet and dry permittivities.

The dielectric loss can be discussed in terms of apparent conductivity or loss tangent. Low frequency conductivity, or d.c. conductivity, is of the order of 10^{-8} to 10^{-7} ohm^{-1} m^{-1} for very pure ice around -10°C, and it decreases with decreasing temperature at a few percent per degree, corresponding to an activation energy of 0.3 to 0.4 eV. Above the Debye dispersion frequency, apparent a.c. conductivity increases to about 10^{-5} ohm^{-1} m^{-1} at -10°C and stays at that value up to a frequency of 10^8 Hz. For this high frequency range, the loss factor and loss tangent are

inversely proportional to frequency. Above 10^8 Hz, apparent
conductivity again increases, exceeding 10^{-4} ohm^{-1} m^{-1} at 10^{10}
Hz and higher frequencies.

The d.c. conductivity of very clean water at 0^oC is
about 10^{-5} to 10^{-4} ohm^{-1} m^{-1}, so that conductivity provides
a sensitive means of detecting liquid water in snow.

Another thing that might be mentioned is the electrical
charge carried on particles of falling and blowing snow.
Charges vary in size and magnitude from particle to particle,
and overall imbalance can affect the local atmospheric
potential gradient. Charges can be transferred to structures,
antennas, aircraft and so on, leading to corona discharge at
points of highest charge density, and to radio noise in some
frequency bands. There are a number of mechanisms that can
produce charging of ice particles, but their relative
significances under varying atmospheric conditions are not
well established.

One of the most interesting charging mechanisms is
based on the so-called thermoelectric effect. When a
temperature gradient exists across prisms or particles of
ice, a corresponding electrical potential gradient develops,
with the warm end negative and the cold end positive.
Below -10^oC, potential difference is approximately proportional
to temperature difference, at about 1.5 to 3.5 mV $^oC^{-1}$.
Brief contact between ice particles of different temperature
creates a potential difference and charge transfer. Maximum
potential difference varies with contact duration, peaking
after about 7 ms. It also increases with impact velocity in
the 0.1 m s^{-1} range.

Discussion and Conclusion

This general introduction to ice does not even begin to
treat the basic problems of glaciology, atmospheric physics,
ice physics, geotechnology, and so on. Nevertheless, it
probably suffices to show that ice and snow are quite
complicated materials that deserve a certain amount of
professional respect.

Although there is still a lot to be learned about the
various kinds of ice, a great deal is already known. Plenty
of information is available, and field research projects,
especially high altitude studies, can be run more efficiently
if existing data are used to the full. However, it is not
always easy to acquire information on ice when working
outside major research centers.

A computerized information retrieval system for ice and cold regions science does not yet exist, but there is a "data base" in the *Bibliography on Cold Regions Science and Technology*. This bibliography, which runs to 30 volumes so far, is compiled on a continuing basis by the Library of Congress for the U.S. Army Cold Regions Research and Engineering Laboratory (CRREL). Numerous CRREL publications deal with ice and snow. Among major journals, the *Journal of Glaciology* covers the physical science aspect, while *Cryobiology* deals with living matter at low temperatures. Relevant foreign language journals are published in Japan, Germany, and Norway, and there are numerous publications dealing with snow, ice, and frozen ground in the USSR. Symposia on ice are held periodically by the International Glaciological Society, the International Commission on Snow and Ice, and the International Association for Hydraulic Research. Major international conferences on permafrost are now being held every four years, sponsored mainly by the Academies of Science of the USA and USSR and the National Research Council of Canada. There are a number of books on ice and related topics, the most notable recent one being *Ice Physics* by P.V. Hobbs (1). Some snow and ice research centers publish their own extended series of reports, and maintain specialist libraries.

References

1. P.V. Hobbs, <u>Ice Physics</u>. Oxford University Press, 837 pp. (1974).

6

High Mountain Ecosystems

Evolution, Structure, Operation and Maintenance

W. Dwight Billings

High mountains exist on all continents, on many islands, and in all latitudes (Figure 1). The definition of "high mountain" has been debated at length. In relation to natural biological systems, high elevation alone is insufficient to differentiate such systems in either structure or operation from those nearer sea level. Rather, it is the *total* environment above and beyond climatic timberline which determines and controls true "alpine" ecosystems.

Elevation interacts with latitude as well as with permanent snowline so that relatively low mountains at high latitudes or near stormy middle-latitude coastlines can be "alpine" in a "high-mountain" sense. In contrast, some higher forested mountains in mid-continent may not be "alpine" even though atmospheric pressure is lower. The absence of trees and the presence of at least some late-lying snow during a cold, short summer are absolutely necessary for the maintenance of alpine ecosystems outside the tropics. Within the tropics, however, such systems may have occasional light snowfalls and nightly frost instead of late-lying snow. Thus, in an ecological sense there are many kinds of "high mountain" environments. Because of differences in biological diversity due to a variety of factors, there is an even greater number of alpine ecosystem types.

Evolution of High Mountain Ecosystems

In regard to the evolution of natural ecosystems, there is considerable resemblance between many continental high mountains and oceanic islands. Islands have been buffered by surrounding seas against invasions by plants and animals; high mountains are similarly buffered by "seas" of surrounding lowland deserts, grasslands, or forests. Cold environments, unfavorable to most organisms, provide an additional buffer for high mountains.

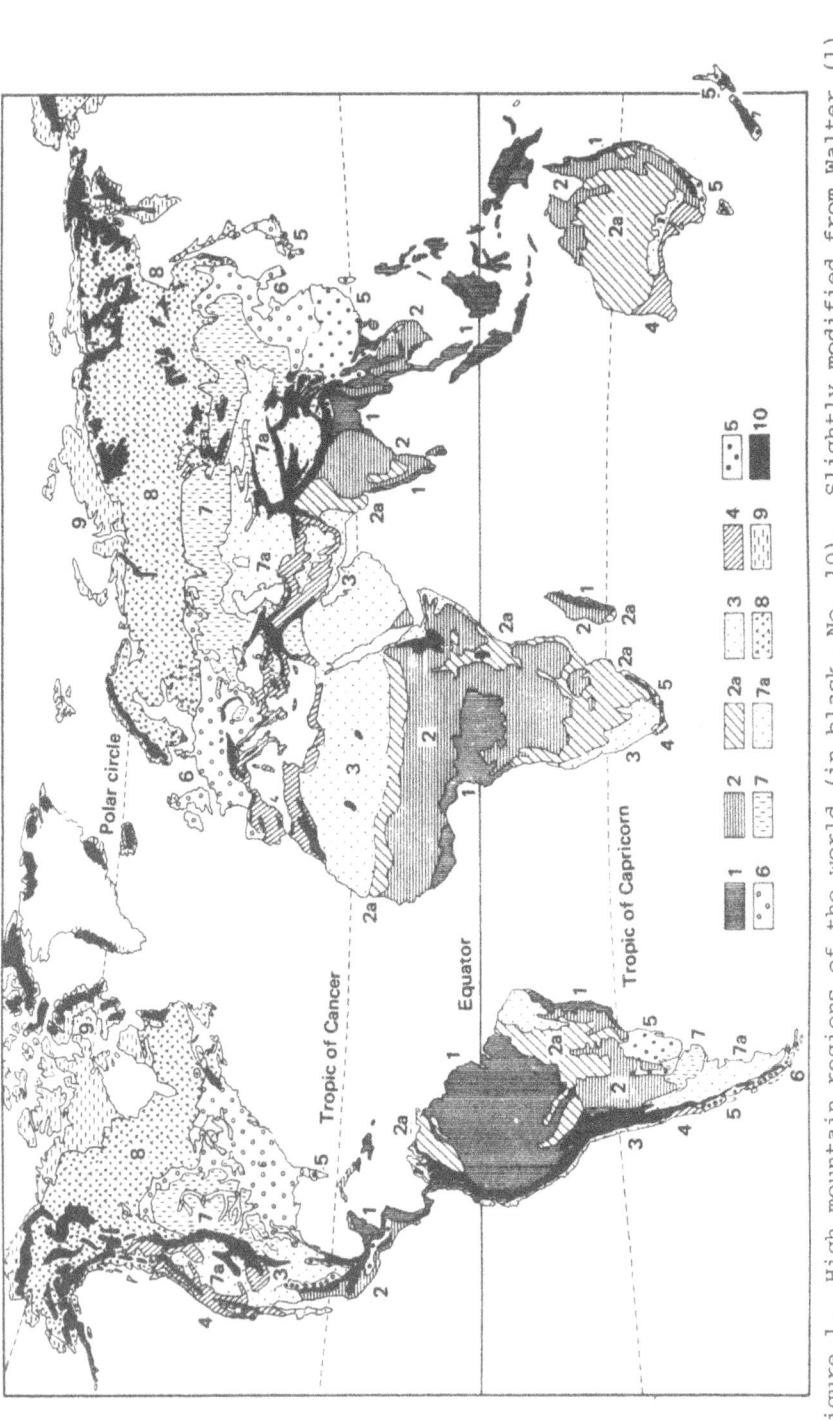

Figure 1. High mountain regions of the world (in black, No. 10). Slightly modified from Walter (1). Reproduced by permission of Springer-Verlag, New York.

Some continental mountains, such as San Francisco Peaks, Arizona, and certain other volcanoes, are relatively isolated and resemble lone islands in the sea. Many mountains, however, occur in ranges or long systems with the higher peaks connected by high ridges or passes. In this way, the higher mountains in such systems resemble islands in an archipelago. The Alps, the Himalaya, and particularly the great American Cordillera extending from Alaska down through the Rocky Mountains and Andes to Tierra del Fuego are montane archipelagos. These relatively continuous systems usually show at least a gradient of ecosystemic relationships along the peaks. This relationship gradient results from the relative nearness of high mountains to each other which makes "island-hopping" by plants and animals somewhat easier than the longer jumps required to isolated mountains. During colder climatic periods, montane "islands" enlarge and may even join; migrations then become somewhat easier. Vuilleumier (2) has demonstrated this for birds in the páramo "islands" of the northern Andes and Simpson (3) has shown the same thing for plant species in the páramos. Brown (4) and Billings (5) found similar situations in the Great Basin mountain ranges of western North America for small mammals and plants, respectively. Colder climates enlarge isolated mountain "islands" also. But usually this is not enough to encourage long-distance migration very much, even though a larger target is provided. There are exceptions, however, which are difficult to explain. San Francisco Peaks, small in area and isolated as they are, have an alpine flora strongly related to that of the Rocky Mountains. Climatic fluctuations on isolated mountains does encourage ecosystem enrichment by evolution and upward migration of biota from the lower slopes and adjacent lowlands (6).

Extreme cases in the evolution of alpine and montane ecosystems occur on high mountains on oceanic islands. Good examples are Haleakala on Maui, Mauna Kea and Mauna Loa on Hawaii, and the high peak on Île de la Possession in the Crozet Islands and that of Kerguelen. Such peaks are buffered not only by long, sterile, over-ocean distances but by cold high-mountain environments. This is particularly true of the mountains on subantarctic Kerguelen and Île de la Possession but is also evident on the high Hawaiian volcanoes. In such cases, those biota which happen to reach these island mountains receive a double dose of extreme isolation. The result is a scanty and/or highly endemic biota consisting of a few closely related but distinct taxa derived by evolution and adaptive radiation among the plants. Such alpine ecosystems appear deceptively simple. The plant components consist of only a few genera and species, often closely related. Mammals are lacking because of the inability to migrate to the island, but birds are an integral part of such systems.

Long ago, a tarweed from western North America established a beachhead on one of the Hawaiian islands. As its offspring migrated up the mountains, they evolved into at least three new genera: *Dubautia*, *Wilkesia*, and *Argyroxiphium* (7). On the dry, cold, cindery crater rim on the southwest side of Haleakala on Maui, one can observe a new high mountain ecosystem evolving. Here, at elevations above about 2200 m and extending to about 3050 m, *Dubautia* has joined the shrubby descendents of plant immigrants from the southwest Pacific to form a chaparral-like community dominated by a few species of brittle shrubs. Near the dry, cold rim itself, the other shrubs drop out leaving only scattered low shrubs of *Dubautia menziesii* and its close relative, the shiny-leaved giant rosettes of *Argyroxiphium sandwicense*, "ahinahina" or "silversword." The double isolation provided by the cold, bright alpine environment and long oceanic distances have resulted in the evolution of a relatively simple biological community, and a unique one. The biosystematic work of Carlquist (7) and others has made it possible to trace the genetics of its plant origins back to the coast of California more than 4000 km to the northeast--and only open ocean between.

The insects in the high Haleakala ecosystem show similar rapid and specific evolution. There is a tendency toward loss of ability to fly. One species of lacewing, *Pseudospectra cookeorum* (Neuroptera) is endemic to this ecosystem and occurs on shrubs of *Dubautia menziesii* just inside the crater rim of Haleakala (7). Among the few bird taxa in the shrubby Haleakala ecosystem is the 'i'iwi (*Vestiara coccinea*), a member of the endemic Hawaiian honeycreeper family (Drepaniidae). As Amadon (8) has shown, this family has gone through relatively rapid adaptive radiation since its warbler-like or tanager-like ancestors reached these islands. There are a number of genera (some extinct) in the family. Most are forest birds, including 'i'iwi. But 'i'iwi does get into the chaparral-like subalpine Haleakala ecosystem where it feeds primarily on nectar of some flowers but may also prey upon insects. The endemic Maui creeper, 'alauahio, (*Loxops maculata newtoni*), has evolved into a true predator of insects and spiders in the bark of shrubs in the subalpine and alpine ecosystems on Haleakala (9, 10).

Haleakala provides an isolated lava and cinder substratum which extends into the bright, dry atmosphere above the trade-wind inversion. Plant, bird, insect, and other immigrants have arrived, evolved, and aggregated into the biotic part of a new and unique high mountain ecosystem.

Admittedly, this Haleakala system is a bit different from most alpine ecosystems in its isolation, relative simplicity, and recent age. But, it could be used as a simple model of the evolution of a high mountain ecosystem if we knew it better than we do. Unfortunately, two organisms which are recent immigrants to the system and which did not evolve with it may make this model impossible to construct; they are people and goats.

As with the Haleakala alpine ecosystem, all high mountain ecosystems are at the upper ends of environmental and evolutionary gradients that originate in the surrounding lowlands. On young mountains such as Haleakala, the Sierra Nevada of California, and the northern Andes, these gradients are being stretched at their upper ends as the mountains continue to rise. New low-temperature, low-pressure, and high-radiation environments on such mountains provide new alpine habitats for preadapted migrating organisms. This seems to be true on all new mountains whether tropical, middle latitude, or subpolar. Atmospheric climatic changes superimposed on this mountain-building process result in fluctuations in such ecosystem evolution--but the process continues. On older mountain ranges, such as the Appalachians, new alpine ecosystems, if present at all, are more likely to be disharmonic and not the result of much evolution from below. Such new systems have resulted from the migration of taxa upward and poleward due to warming of climates during Pleistocene interglacials. Many of these taxa, having evolved in subpolar or distant alpine environments, are already adapted to low summer temperatures; they migrated southward in glacial times. The alpine ecosystem of Mt. Washington in the northern Appalachians is a new ecosystem but most of its species are relatively old and of arctic derivation; they just happened to be nearby during deglaciation of the mountain in post-Wisconsin times.

The biological component of an alpine ecosystem is, of course, dependent upon the primary productivity of its green plants. This productivity results from the interaction of genetically controlled photosynthetic processes and the cool mountain environment. Apparently, the evolution of photosynthetic systems adapted to cold summers is a rather slow process and a relatively rare thing. For example, the number of species of vascular plants decreases steeply from tropical moist regions toward the poles. Most of this decrease occurs poleward of the middle-latitudes where warm summers prevail at low elevations. In such warm summer climates, there are some dominant species in herbaceous communities which have the C_4 photosynthetic mode which results in relatively high productivity. As of this time,

this photosynthetic system has not been found in subpolar
and polar species. All such species which have been examined
at high latitudes have the C_3 system (11). Up to the present
time, this seems to hold true for all plants above alpine
timberline although a C_4 *Muhlenbergia* has been found in
subalpine vegetation in Wyoming (12).

Data on species numbers with increasing elevation
should show a decrease comparable to that with increasing
latitude. However, accurate floristic data from elevational
gradients on a single mountain range are not always easy to
find even though in many places the floras are known and the
techniques are simple but time consuming. Reisigl and
Pitschmann (13) have done this for the Ötztaler Alps in the
Tyrol and Breckle (14) has some fine data from the Hindu
Kush of eastern Afghanistan. In the Ötztaler Alps, 105
species of vascular plants occur above 3000 m, 39 above 3500 m,
and only 8 above 4000 m. In comparison, the Hindu Kush,
higher in elevation and rising from a higher platform, has
377 species above 4000 m, 162 above 4500 m, and 36 above
5000 m. Of those above 4000 m, about 82 species or 25% are
endemic to the Hindu Kush while only 10% of those above 5000 m
are endemic. Of all species above 4000 m, only about 15 to
17% occur in the Arctic. This latter figure is about the
same percentage as one finds in the Sierra Nevada of California,
and seems to be typical of relatively young mountain ranges
or those with high elevation connection to the Arctic.
Endemism is probably much higher in the northern and central
Andes, and the degree of relationship to the Arctic is
certainly even lower. Unfortunately, the rich alpine floras
of the Andes are not yet well enough known to prove this;
but those data which do exist indicate that both estimates
are probably true.

In the large mountain systems, such as the American
Cordillera, once plant and animal taxa have migrated up into
the high mountains and evolved adaptations to the alpine
environment, many of the species are able to migrate and
join other high mountain ecosystems (15). Such migrations
may take place overland during cold periods when timberlines
are low. Or, species can "island-hop" in short or long
jumps. The result of such migrations in the large and
connected mountain aggregations or cordilleras is the
development of extensive alpine ecosystem types which resemble
each other ín composition and operation even though separated
by considerable distance. This is more likely to be so on
older and/or less isolated mountains.

A final point in regard to evolution of high mountain
ecosystems, including their soils, is that almost all have

evolved in the absence of people. Only crop and pasture
ecosystems, such as those of the central Andes, are exceptions.
Many also evolved in the absence of hooved animals. Those
on islands such as New Zealand and Hawaii, for example, had
neither people nor hooved herbivores until recent centuries.
Such sudden introductions have proved to be disastrous in
almost every case. Individual organisms are not fragile in
the face of severe physical environments to which they have
become adapted through evolution. But the whole alpine
ecosystem *is* fragile when presented with aggressive organisms
which severely damage or destroy the vegetation and, ultimately,
the shallow mountain soils.

Structural and Operational
Characteristics of Alpine Ecosystems

Environment

Since mountain tops vary in elevation and latitude, it
is not surprising that, in spite of sharing many environmental
characteristics in common, their environments do differ.
These differences are more easily seen along gradients of
various kinds and lengths. The most obvious of these gradients
are latitudinal (from equatorial to polar regions) and
elevational (from the lower slopes to the summits). These
"macrogradients" have a great deal of influence on the
amounts of solar radiation received, the length of day,
temperatures, and amounts of precipitation. As one result,
such gradients have considerable control of the kinds of
plants and animals which are available in the region to make
up the biological parts of mountain ecosystems.

Within these large gradients are smaller environmental
gradients which exist on every mountain. These, along with
bedrock differences, control almost all the local distribution
patterns of plant and animal species within the constraints
set by the elevational gradient on that particular mountain.
The smaller gradients are determined primarily by topography.
Topography, in itself, does not control growth and distribution
of the alpine organisms but does moderate those factors
which interact directly with the organisms: solar radiation,
soil moisture, soil and air temperatures, wind, and both the
blasting and protecting aspects of snow. Ridges, slopes,
valley and cirque floors act together along what I have
called a "mesotopographic" gradient (16) as diagramed in
Figure 2.

A mesotopographic gradient crosses a ridge top and
descends the lee slope to a small valley or to a cirque
floor in a distance of 50 to a few hundred meters or so.

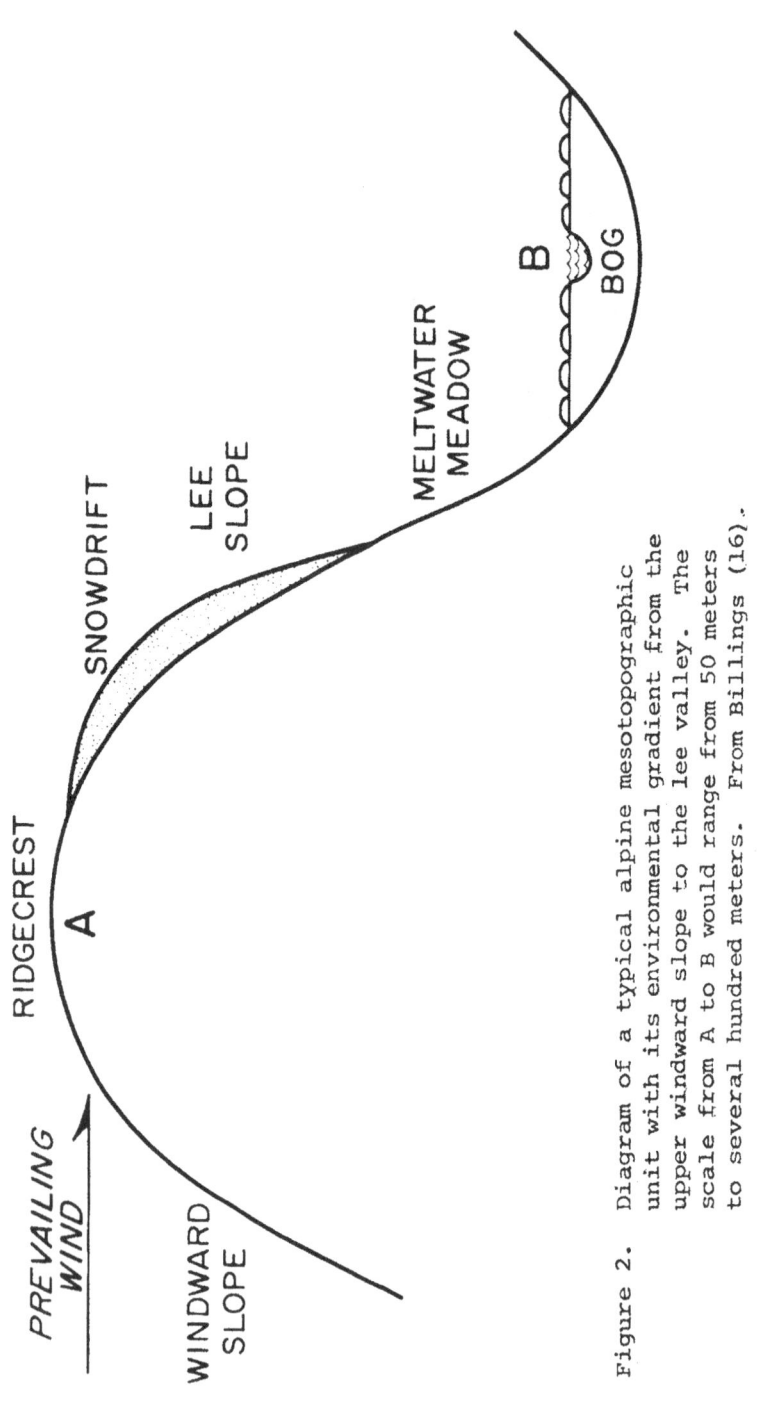

Figure 2. Diagram of a typical alpine mesotopographic unit with its environmental gradient from the upper windward slope to the lee valley. The scale from A to B would range from 50 meters to several hundred meters. From Billings (16).

Ridge tops outside the tropics are exposed to strong winds and are relatively cold and dry. Snow is blown into drifts on the upper part of the lee slope. Such drifts do not begin melting until the following summer; some do not disappear except in very dry years, if then. Meltwater from the drifts keeps the lower lee slopes moist; this water eventually accumulates in small bogs, tarns, and brooks. A mesotopographic gradient strongly modifies the energy-determined latitudinal and elevational gradients by shifting the snow patterns and, ultimately, the amount and timing of water available for plant growth. The result is a steep vegetational gradient in which analogs of vegetation along the whole latitudinal gradient occur in close proximity. In tropical mountains, wind speeds strong enough to produce large snowdrifts occur only at very high elevations where the number of plant species is low. Therefore, in the tropics, such gradients affect the distribution of alpine vegetation relatively weakly and over smaller areas.

Within mesotopographic gradients, boulders or rock outcrops produce small, elongated snowdrifts in their lee. Small depressions also catch blowing snow. Even these small drifts produce small environmental gradients and the captured snow, because of the winter protection it affords, is reflected in vegetational patches consisting of species needing such protection. Anything of this nature can result in a microtopographic gradient.

Both mesotopographic and microtopographic gradients result in rather orderly patterns in high mountain environments. These environmental patterns result in vegetational and biological community patterns to match the environmental patterns. I find it convenient to study high mountain ecosystems by utilizing a series of mesotopographic units on a single mountain. The combination of a mesotopographic unit with its microtopographic units and variations in bedrock produces a certain patchiness in high mountain environments. Because of the gradients, there is some orderliness to this patchiness. Orderly or not, patchiness if well-expressed results in habitat diversity and greater biological diversity than might be expected.

In high mountain environments, it is the physical factors which dominate and control the ecosystem. The plants and animals are there in patterns determined by local gradients in solar ultraviolet and visible radiation, net radiation, low soil and air temperatures, snow distribution, wind speeds, length of "growing" or snow-free season, steepness of slope, type of rock, and soil characteristics including soil frost action. Plants are small in these

cold, windy environments; such vegetation cannot modify or
"soften" the environment as does a montane forest lower down
the slope.   Consequently, the biological portion of the
environment seems to be less important than physical factors
in determining patterns and growth rates in alpine biological
communities.   However, this all may be a bit deceptive.   We
know that large native herbivores have effects on the vegetation.
And, the introduction of hooved herbivores into alpine
ecosystems on islands wreaks havoc with plants and soil.   In
some alpine ecosystems, insect predation can be very intense
on both vegetative and reproductive systems of the plants.
In this regard, the puna of the central Peruvian Andes comes
to mind.   Perhaps we just do not know alpine environments
and ecosystems well enough yet--even in their relatively
natural states.   When one adds people to the picture, the
biological part of the alpine environment becomes dominant
indeed, and destruction of these high mountain ecosystems
can and does occur.

Vegetation and Primary Productivity

As in all ecosystems, the transfer of operational
energy from the environment to the biological community is
by photosynthesis in green plants.   In alpine systems, these
plants are small and often do not completely cover the
ground; in fact, in those communities at the highest elevations,
rock and snow reduce vegetational cover to a few scattered
plants.   However, high mountain vegetation, in spite of its
low stature in a low temperature environment, captures and
stores as much or more energy per growing season day than
does the vegetation in other grasslands and even in certain
croplands (17).   However, because of the short alpine growing
season, energy capture on an annual basis is somewhat less
than in these other herbaceous vegetations which operate
through a longer growing season at lower elevations.

Plant Life Forms.   Outside of the tropical mountains,
alpine vegetation is dominated on the better soils by perennial
graminoids (grasses and sedges).   This is true also in some
equatorial and tropical mountain ecosystems such as the
páramos of central and southern Ecuador and in much of the
Peruvian puna.   Perennial dicotyledonous plants usually grow
with the graminoids; most of these are herbaceous but there
are also some depressed shrubs and semishrubs.   On rocky
outcrops and very shallow soils, the principal life forms
are usually perennial herbs or cushion plants.

As in the Arctic, alpine vegetation consists almost
entirely of perennial plants.   Almost all perennial graminoids
have much more living tissue belowground than· above; most,

but not all, dicotyledons also have most of their biomass within the soil. Annual species are absent or very few in number. These annual plants are miniature, and they contribute little to vegetational cover or productivity. The following list of plant life forms is arranged in approximate order of importance in high mountain vegetation outside the tropics.

Perennials
    Graminoids
    Herbaceous dicotyledons
        Single-stemmed dicotyledons
        Cushion plants
        Small basal rosette plants
    Dwarf shrubs, either evergreen or deciduous
    Mosses
    Lichens
Annual graminoids or dicotyledons

Many tropical mountains exhibit certain life forms which are missing or very rare in the alpine regions of the middle latitudes. The most striking of these life forms is the giant rosette plant. This form has evolved independently on the high mountains of equatorial east Africa, in the northern and central Andes, and on the mountains of Maui and Hawaii. Most of these plants are single stemmed while some resemble their branched ancestral forms. Others are relatively stemless and sit on the ground as large silvery-leaved bowls reflecting light inward onto the younger leaves and bud. Among the genera in which this life form has evolved are *Senecio*, *Lobelia*, *Puya*, *Lupinus*, *Espeletia*, and *Argyroxiphium*.

A second unusual life form in high mountains is that of the succulent cactus stem. Large white mats of the white-hairy cylindrical *Opuntia floccosa* occur as high as 4200 m in the puna of Peru accompanied here and there on limestone by a true, but small, barrel cactus of the endemic genus *Oroya*. These latter reach a height of at least 4 dm and a diameter of 2 to 3 dm. Farther south on the east side of the Andes, even larger cacti occur on fairly high mountains where Bolivia, Argentina, and Chile come together. The cactus life form does not occur in the truly alpine areas of the western United States but a semiburied small ball-like cactus, *Pediocactus simpsonii*, reaches elevations of almost 3000 m on dry subalpine ridges in the Rocky Mountains.

Vegetational Structure. In spite of the variety of plant life forms, most alpine vegetation is dominated by perennial graminoids. Associated with these grasses and sedges are herbaceous dicotyledons and dwarf shrubs. The vegetational gradient at any given elevation is closely

correlated with and controlled by the snow and water available
in different segments of the mesotopographic unit. The most
productive and luxuriant alpine vegetation lies in the
meltwater meadows below melting snowdrifts. The least
productive and most sparse vegetation lies under the snowdrift
itself and is exposed only in late summer. Here, a few
dwarf perennial dicotyledons are the only plants scattered
over barren rocks, gravel, and a bit of soil.

Except for lists of flora, more research has been done
on vegetational structure in high mountains than on any
other part of alpine ecosystems. Rather than list so many
sources of vegetational information here, it is easier to
refer to the literature lists in several review papers (18,
19, 20, 21, 22).

Plant Processes. Alpine plants in the mountains of the
middle latitudes and polar regions are dormant for 9 to 10
months of the year. During this time, temperatures of the
surrounding air and soil are at or below freezing. The
plants are snow covered except on the wind-swept ridges
where snow cover is sporadic. A number of kinds of plants
on these ridges or upper windward slopes actually cannot
tolerate protective snow cover and occur only where high
winds of winter keep the ground bare and frozen. Among
these are a sedge, *Kobresia*, and certain small rosette
saxifrages and cushion plants. Such plants, in dormant
condition, are adapted to environmental conditions which
seem as severe as any on earth: extremely cold and dry
because of exposure. The perennating or live parts of
alpine plants in long-winter mountains are in the buds,
roots, rhizomes; in evergreen plants, the younger leaves
also remain alive. Some plants in the high tropical mountains
also have a dormant season but most do not since one day
there is just about like any other throughout the year.
Except for a short dry season in some of these tropical
mountains, a mild winter comes every night and almost every
day is a "growing season" in which some metabolic activity
and even growth occurs. In the following paragraphs, very
brief descriptions are given of plant processes as they occur
during the short summers on high mountains outside of the
tropics. More information is available in other publications
(18, 19, 20).

The perennial herbaceous life forms of alpine plants
enable the vegetation to break dormancy rather rapidly after
snowmelt or after soil thaw on snow-free ridges. This usually
occurs about the time of the summer solstice but is a bit
later in the areas around slowly melting snowdrifts. Rapid
growth of shoots and leaves is accomplished by translocation

of carbohydrates and lipids from roots, rhizomes, and old evergreen leaves. Respiration rates are high at this time which speeds up most metabolic processes. Photosynthesis, however, starts out slowly as compared to respiration due partly to the time it takes for chlorophyll synthesis. The result is net dry weight loss for the first week or so after snow release.

Positive net photosynthesis takes over during the daylight hours as soon as the plants are green with enough chlorophyll. The photosynthetic pathway apparently is C_3 in all alpine graminoids and probably for almost all dicotyledons also. I do not know of any truly alpine graminoids which have the C_4 photosynthetic mode. Harrison (12), however, has found a C_4 *Muhlenbergia* in the subalpine zone of the Laramie Mountains in Wyoming. Teeri and Stowe (11), using multiple regression with 32 grass floras in North America and 19 environmental variables or combinations of such variables, found that mean daily minimum temperature in July was the best predictor of the percentage of C_4 graminoid species in a flora. No C_4 species occurred in those places where the mean daily minimum was below 8°C. In a linear regression of actual percentage of C_4 grass species against "normal" mean daily minimum temperature in July, the percentage of C_4 species becomes zero at about 7°C. The correlation coefficient (r) for this regression was .972. Since temperatures in middle latitude alpine areas go well below 7°C almost every night, this pretty well rules out the presence of C_4 graminoid species. Admittedly, more basic research is needed on this point. But even if a C_4 species is found in an alpine area, it would likely be so rare as to have almost no effect on the productivity of alpine vegetation. Alpine areas in tropical mountains also get cold at night; if graminoids there perform as the C_4 species in North America do, it is unlikely that high Andean grasses, for example, would operate in the C_4 pathway. For alpine dicotyledons, there is little or no information on the presence of C_4 species. It seems probable that alpine dicotyledons are almost entirely C_3 species also. There is a possibility that C_4 dicotyledons might be found in alpine regions of high mountains which are in or near deserts.

There are species of *Sedum* in temperate zone alpine areas and of *Echeverria* in the páramos and punas of the Andes. These genera are members of the Crassulaceae. They could be suspected of having Crassulacean Acid Metabolism (CAM) in which the stomates are closed during the day and open at night. In this photosynthetic mode, carbon dioxide enters the leaf during the night and the carbon is stored in an organic acid, usually malic acid, in the mesophyll cells. During daylight hours, this malic acid is broken down and the

carbon fixed in the light reaction of the C_3 pathway. There
is a possibility that certain alpine members of the Crassulaceae
may operate facultatively in the CAM mode during very dry
periods (23). *Opuntia floccosa* and *Oroya*, cacti of the
Peruvian puna, seem likely candidates for CAM photosynthesis
during the annual dry season there. But as yet, there is no
information on this.

Since almost all alpine plants appear to be C_3 plants,
this means that some organic compounds are lost in photorespiration in addition to those used in dark respiration. We
know that dark respiration rates of alpine plants are relatively
high at low temperatures but we know little or nothing about
rates of photorespiration in alpine plants. If high, photorespiration could be involved in decreasing net photosynthesis
rates and, thus, net primary productivity in alpine ecosystems.

Alpine weather is known for its rapid changes in temperature and light as sunshine is replaced by storm clouds. At
least some alpine species show the ability to adjust photosynthetic and dark respiration rates rather quickly to such
environmental changes. Alpine ecotypes of *Oxyria digyna*
(alpine sorrel) have shown "ideal" homeostasis in adjusting
net photosynthetic rates to temperature changes while arctic
ecotypes of the same species in the, same experiment showed
only a low degree of partial homeostatic adjustment (24).
Following this acclimation down to the level of the organelle,
isolated chloroplasts from plants of alpine *Oxyria* ecotypes
showed an increased ability to adjust the Hill reaction rate
to higher values with increasing light intensity when grown
at low temperatures than when grown at higher temperatures
(24). Arctic ecotypes showed some ability to acclimate in
similar fashion, but more slowly, and to not the same extent.
Arctic ecotypes required continuous photoperiod to adjust
even this much in the Hill reaction while alpine ecotypes
acclimated easily at both long and short photoperiods.
Alpine plants of the grass, *Deschampsia caespitosa*, also
showed greater flexibility in adjusting their Hill reaction
rates to temperature changes than did arctic ecotypes of the
same species (25).

Dark respiration rates of whole plants of 17 arctic and
alpine ecotypes of *Oxyria* showed no apparent differences in
temperature acclimation among ecotypes; homeostasis was
almost "ideal" to temperature changes in all ecotypes (24).
However, alpine populations, in general, showed less respir-
ational loss of carbon when grown at warm temperatures than
did plants from arctic populations. Again, following this
process down to the organelle level, arctic ecotypes showed
much higher mitochondrial oxidation rates when grown at low

temperatures than did alpine ecotypes grown at the same low temperatures. However, when plants were grown at warm temperatures, mitochondrial oxidation rates were much lower than in those grown at cold temperatures; there was no difference between ecotypes (24). The rate change, therefore, was less in the alpine ecotypes as compared to that in the arctic ecotypes. This is another example of alpine ecotypes, as compared to arctic ecotypes, showing a greater ability to adapt by making rapid changes in metabolic rates in the face of wide temperature changes.

Photosynthesis rates of Northern Hemisphere alpine plants as measured in the field are difficult to interpret since different research projects may use different techniques and express the CO_2 uptake on different bases. In the Alps and in the Rocky Mountains, maximum rates between 4 and 13.5 mg $CO_2 \cdot dm^{-2}$(two surfaces)$\cdot hr^{-1}$ have been reported for several species (26, 27). In these studies, whole plants were used; average maximum values were around 10 mg $CO_2 \cdot dm^{-2} \cdot hr^{-1}$. In other studies, single leaves often have been used, different techniques have been used (infrared gas analyzer, $^{14}CO_2$, etc.), and CO_2 uptake has been expressed on square decimeter, grams dry weight, and grams fresh weight bases. These different approaches have made it difficult to know just what field photosynthesis rates mean and how comparable they are. For example, it is known that some species will have apparent net photosynthesis rates higher than others if expressed on a leaf area basis and lower than the others if expressed on a dry weight basis. Standardization is needed both in field and laboratory techniques for measuring photosynthesis and respiration.

The photosynthate produced by high mountain plants is used not only in current growth and reproduction but also as storage compounds in roots, rhizomes, and evergreen leaves. A very large proportion is stored in this way, particularly late in the growing season, and is utilized in early growth the following year (28). In most alpine plants, the storage compounds are carbohydrates (28) but lipids are also stored, particularly in small evergreen rosette plants and in the evergreen leaves of low shrubs such as *Ledum groenlandicum* and *Diapensia lapponica* (29).

The translocation of carbon from starch photosynthate in the chloroplasts to other parts of the plant at low temperatures appears to be one of the adaptations of alpine plants. This degradation of starch to sucrose or its possible change to osmiophilic globules (probably lipids) in the chloroplasts of *Oxyria digyna* takes place during cold nights (6). The same process of starch degradation to sucrose takes place in leaves of *Encelia virginensis*, a species of the warmer lower slopes at higher temperatures (20-23°C). However, if *Encelia* plants are exposed to the low temperature nights (8-11°C) in which *Oxyria* successfully degrades starch to sucrose, the starch grains still remain in the *Encelia* chloroplasts the next morning. Such starch accumulation not only reduces photosynthesis but also probably leads to the breakdown of the chloroplasts themselves. This ability to degrade starch to sucrose at low temperatures and to translocate the sucrose to storage organs at the same low temperatures may be one of the principal steps in the evolution of adaptations of plants to high mountain or polar environments. It is probable that the great majority of plants cannot do this. It would appear that such degradation and translocation can occur in certain crop plants such as potatoes but not in others. Much more research on this apparent adaptation is needed; it would be particularly helpful in high mountain agriculture.

Flowering in high mountain plants takes place almost entirely from flower bud primordia formed the year before or even earlier. Such preformed flower buds enable alpine plants to bloom rather soon after breaking of dormancy. If the season continues favorably, fruit and seed set can occur in late summer. Many of these plants also can reproduce vegetatively by bulbils, rhizomes, stolons, and bulbs. All reproductive structures are rich in carbohydrates, lipids, and, to some extent, proteins. As such, they are important parts of primary production in the ecosystem and provide considerable food for higher tropic levels.

Primary Productivity. Primary productivity in an ecosyste is the rate at which solar energy is captured by the vegetation per unit area of land surface or water volume. Net primary productivity is essentially equal to the rate of photosynthesis minus the respiratory costs of carrying on growth, photosynthesis, and synthesis of other compounds. In theory, it is possible to measure primary productivity over a growing season by measuring net photosynthesis (gross photosynthesis minus respiration) with an infrared gas analyzer which measures carbon dioxide uptake and output from single plants or small plots of vegetation enclosed in a transparent and temperature-regulated chamber or cuvette. This ideal method is very delicate, and expensive in equipment and technicians. It is also subject to component failures in inconvenient places. It requires electrical power. Its use is confined to places with such a power supply or to places which can be reached by a mobile laboratory containing a power source and all equipment This technique can be and has been used in high mountain environments to measure photosynthesis and primary productivity (27, 30, 31, 32). Excellent measurements of photosynthesis

in alpine plants *in situ* in the Tyrolean Alps have been made
continuously (as long as 1 month) by Moser (31, 32) on an
exposed ridge ("Hoher Nebelkogel") using an infrared analyzer
with electrical power supplied by a wind-operated generator.
These measurements show some positive net photosynthesis at
leaf temperatures as low as -6°C (31, 32).

Since the measurement of primary productivity in alpine
ecosystems is expensive and fraught with difficulties, the
usual method of measurement is the harvest method. This
consists simply of clipping new growth per unit area of
alpine vegetation at certain time intervals through the
season, or at the end of the season. Because of vegetational
variation, enough sample plots must be harvested to make the
results statistically valid. There is another difficulty
with using the harvest method on alpine vegetation. This
arises because most of the living material in alpine vegetation
is underground in roots, rhizomes, tubers, and bulbs. These
are not only difficult to extricate completely but it is
difficult to distinguish between live and dead tissue.
Therefore, most harvest method measurements of primary
productivity in high mountains have been based only on
aboveground material. Some measurements have been made by
actually measuring or estimating belowground growth. Since
animals eat rhizomes, bulbs, and tubers in mountain ecosystems,
productivity rates are not really very accurate in alpine
ecosystems unless belowground production is at least estimated.
And, in the Andes, many people are supported by tubers
(potatoes, *Oxalis*, *Ullucus*, etc.) derived from high mountain
crop ecosystems.

In the harvest method, the fresh material is oven-dried,
weighed, and the grams of oven-dry material expressed on a
unit area basis per unit time. Measurements are more useful
in understanding energy capture and flow if calorific contents
of the dried material are determined in a bomb calorimeter.

On an annual basis, productivity in alpine vegetation is
quite low compared to that of the rest of the biosphere; only
extreme deserts (including polar deserts) are lower. This is
because alpine areas are only productive 10 to 25% of the
year, and under low temperatures and occasional drought
stress at that. In terms of average shoot production alone
on Mt. Washington, New Hampshire, Bliss (33) found primary
productivity ranging from 0.5 to 5.0 $g \cdot m^{-2} \cdot day^{-1}$. Scott and
Billings (17) found rates as high as 11 $g \cdot m^{-2} \cdot day^{-1}$ on moist
sites in the Medicine Bow Mts., Wyoming, when belowground
production was included. Wielgolaski (34), using the harvest
method at Hardangervidda, an alpine site in Norway, also took
into account the belowground tissues. He made the latter

measurements by determining which roots took up $^{14}CO_2$. In a dry alpine meadow, mean daily production was 5.1 $g \cdot m^{-2} \cdot day^{-1}$ during the growing season and in a wet meadow, 8.3 $g \cdot m^{-2} \cdot day^{-1}$. Annual figures were 534 $g \cdot m^{-2} \cdot yr^{-1}$ in the dry meadow and 833 $g \cdot m^{-2} \cdot yr^{-1}$ in the wet meadow. Wielgolaski's daily figures lie between those obtained by Scott and Billings: 1.17 $g \cdot m^{-2} \cdot day^{-1}$ for their xeric site (much drier than Wielgolaski's) and 11.08 $g \cdot m^{-2} \cdot day^{-1}$ for their moist site. The figures are in pretty fair agreement. While these figures are rough and variable, the inclusion of belowground parts gives a better answer than can be obtained from the harvest of aboveground material only.

Productivity data in alpine potato fields in the Andes should be fairly easy to obtain by the harvest method. On a daily basis, productivity in these fields may be greater than that in native Andean ecosystems. Using rough figures on tuber yield and weight of dried "chuño" (native dehydrated potato) in Gade (35), and estimating dry weights of aboveground parts, my rough estimate for peasant-grown potatoes at 3600 m in the Vilcanota Valley of the Peruvian puna is a total plant yield of 275 g dry $wt \cdot m^{-2} \cdot yr^{-1}$ for the growing season of 5 months. This is a productivity of only ca. 2 $g \cdot m^{-2} \cdot day^{-1}$. As in a comparison of productivities between any natural ecosystem and a crop ecosystem in the same region, it must be remembered that tillage and fertilizers can increase primary productivity even in high mountains. Again, using Gade's tuber yield figures for the haciendas where chemical fertilizers and guano are used with hybrid potatoes, I come up with a calculated guess of a dry weight total plant yield of ca. 1600 $g \cdot m^{-2} \cdot yr^{-1}$ or about 11 $g \cdot m^{-2} \cdot day^{-1}$ for the 5 month growing season. These potato figures seem low since they are about those found in xeric to wet meadows in the alpine regions of western North America and Norway. Surely, high Andean potato productivity must be higher than this.

Animals in High Mountain Ecosystems

Animals of many kinds, vertebrates and invertebrates, make up the "grazing" food web of alpine ecosystems. Permanent residents or even temporary visitors during the warmer months are dependent during their stay in the mountains on the production of carbon compounds by alpine vegetation. Only birds flying nonstop over mountains are exempt from this dependence.

In any natural high mountain ecosystem such as that of the Brooks Range in Alaska, the native herbivores make up by far the greatest component of animal life. These include both vertebrates and invertebrates. The Brooks Range vertebrates are entirely mammals and birds; amphibians and reptiles are lacking that far north. The wood frog (*Rana sylvatica*) is the only poikilothermic vertebrate to get even close to arctic mountains (36). Terrestrial herbivorous mammals in the Brooks Range include caribou, Dall sheep, moose, marmot, and many others. There is a great variety of bird genera. Insects are the invertebrates in greatest abundance. Insects and their larvae play an important role in the alpine detrital food web also. Carnivores in the Brooks Range include the grizzly bear (also an omnivore), the wolf, and foxes.

In temperate regions and tropical high mountains, there is a much greater diversity of resident and transient animal life than in an arctic range such as the Brooks. Even in these nonarctic mountains, however, amphibians consist of only a few species in North America, and reptiles are extremely rare. Apparently, there is a more strongly developed herpeto-fauna in certain mountain regions of the Old World (37). Herbivorous insects play an important role in many high mountain ecosystems: the Himalaya, the mountains of western North America, and the Peruvian puna, for example.

Natural high mountain grazing food webs are extremely complex. Such food webs have been studied on the Beartooth Plateau in the central Rocky Mountains (36). Pattie and Verbeek (38, 39) found 22 species of mammals and 13 species of birds to be integral parts of this food web. The situation is further complicated by occasional visits by other native birds and mammals, and certainly by the flocks of domestic sheep which are brought into this alpine region for summer grazing. The roles of animals in alpine ecosystems over the earth is too complex a subject to pursue here. Suffice to say that herbivores, omnivores, and carnivores (both vertebrate and invertebrate) are important parts of high mountain ecosystems, and that most of these are native to the systems. However, some are introduced such as the many kinds of deer in New Zealand and the goats on Haleakala. For more detail, reference can be made to the excellent works by Hoffman (36) for high mountain vertebrates and by Mani (40) concerning alpine insects. For most mountain regions, the roles of invertebrates in alpine ecosystems remains a "black box."

Decomposers in High Mountain Ecosystems

Organisms such as fungi, bacteria, protozoans, and invertebrates of many kinds make up and operate the detrital arm of any ecosystem. Turnover rates of carbon and essential mineral nutrients are controlled largely by these organisms, most of which live in or on the ground. In high mountains, even their taxonomy is not well known let alone their abundance and functional role in ecosystems. Because of low temperatures, the soil being frozen for most of the year, and low primary productivity, one would assume that decomposer activity in alpine ecosystems would be low compared to almost all other ecosystems. Very few data exist on decomposition food webs and rates in regions above timberline. Among the better data for information on decomposers and decomposition rates in alpine ecosystems are those for Hardangervidda in Norway by Wielgolaski and his colleagues of the Fennoscandian International Biological Programme (IBP) Project (41, 42). Estimated annual production of soil invertebrates (of which 90% are decomposers) in Fennoscandian alpine tundra areas averages about 0.5 g dry wt·m^{-2} including nematodes. This is only about 0.10 to 0.15% of the primary production and only about 0.2% of microbial production. It is obvious that in this alpine ecosystem, soil invertebrates probably play only a minor role in decomposition as compared to that of microorganisms such as bacteria and fungi. Even so, both on a biomass and a production basis, invertebrates make up a far larger part of the biological community on Handangervidda than do the vertebrates. When nematodes and earthworms are included, invertebrates have an average biomass of about 2 g dry wt·m^{-2}. For the whole alpine plateau, this would be about 20,000 metric tons of invertebrates or about 10 times the biomass of small rodents in a peak year and about 100 times the wild reindeer biomass. Annual production of fungi and bacteria at Hardangervidda is about one-half that of primary productivity at all sites from very dry lichen heath to wet alpine meadow. The production of microorganisms in the dry heath is about 120 g·m^{-2}·yr^{-1}; in a dry meadow, it is about 320 g·m^{-2}·yr^{-1}; and in the wet meadow 465 g·m^{-2}·yr^{-1}.

Most plant material produced at Hardangervidda goes directly through the decomposer food web rather than the grazing food web; the portion of plant material decomposed is considerably more than 90%. About 40% of the aboveground nutrient-rich herbaceous material is decomposed during the first year after production, but only about 10% of the lichens and bryophytes are decomposed each year. Lack of moisture seems to be the most important environmental factor limiting decomposition.

The same factor, lack of moisture, combined with low temperature, is involved in a striking phenomenon seen near timberline in many high mountains in the western United States. After forest fires just below timberline in the Rocky Mountains, it is not unusual for the dead tree trunks to remain standing for 1 to 3 centuries (43). Even after falling, decomposition of these trunks and branches may take several more centuries. An extreme case of slow decomposition

under high mountain conditions can be seen in the White
Mountains of eastern California and western Nevada. Above
the present timberline of *Pinus longaeva* where some living
trees are over 4200 yr old, there are dead trunks of the
same species lying or standing in alpine vegetation at about
3500 m. Dating of the outer wood of these logs by ^{14}C and
ring measurements indicates that the trees died from about
800 to 4000 years ago presumably because of a change to
cooler and more severe climate (44). One tree *still standing*
died about 2400 yr before present. In high mountains as
cold and dry as the White Mountains (one of the Great Basin
ranges), alpine decomposition rates can be slow indeed!

Sensitivity of High Mountain Ecosystems to Environmental Changes

Changes in the Physical Environment

High mountain ecosystems exist in what might be termed
"severe" environments as seen from a human viewpoint.
However, the organisms making up alpine biological communities
are well adapted to such climatic "severity" and also to
short-term changes such as unusually cold winters or summers,
and even to dry years when snowfall is much below normal.
Alpine organisms have evolved adaptations to such a physical
environment or they would not have gotten there, or remained.
The presence of alpine ecotypes in plants, for example,
indicates that such evolution still continues. In certain
kinds of organisms, relatively few in number, evolution of
physiological adaptations may be more rapid than one would
suppose. Such organisms also appear to have the genetic
capability of adapting phenotypically in both structure and
metabolism by acclimatization.

Many high mountain species can migrate up or down the
mountain as climates change. Some are able to find refugia
on certain rock types or in protected microenvironments
either warm or cold, wet or dry. As mountains rise or
climates change, those species which cannot evolve adaptations
to high mountain environments or migrate to local refugia
become extinct, at least locally. If the range of a species
is already large, it may survive elsewhere as part of
another kind of ecosystem; or it may even survive elsewhere
as part of an alpine ecosystem if a few individuals can
evolve adaptations to low temperature growing seasons.
Evolution of this type does not have to occur evenly throughout
the entire range of a species.

Soil formation goes hand-in-hand with evolution of
plants and the development of vegetation. High mountain

soils, once formed, are protected by the native vegetation;
and the native vegetation, in turn, is dependent upon these
soils for water, nutrients, and stability. Under natural
conditions, only glaciation or some other natural, but rare,
catastrophe such as an earthquake, avalanche, or mudslide
may destroy the vegetation and the soil. However, high
mountain soils are often on steep slopes and most of these
soils are very shallow. As such, they are very susceptible
to erosion by water (unusually heavy rainfalls) or even by
wind if the vegetation is injured or dies.

Changes in the Biological Environment

Alpine ecosystems are well adapted to the presence of
their own native organisms including herbivores; each has
its niche. The large grazers are generally somewhat migratory
on both seasonal and space bases. Native insect predators
seldom become epidemic; they maintain a balance with the
primary producers. These green plants not only supply food
but they also hold and help build the mountain soils.

It is the introduction of *new* organisms which throw
alpine ecosystems out of balance. Of these new organisms,
people are the most important and potentially the most
likely to influence such systems to change. This is because
high mountain ecosystems evolved without man as a component.
Primitive man was only a casual visitor and did little to
affect the operation of these potentially unstable systems.
Only in a few places and after long and cautious residence by
fairly primitive agricultural peoples have mountain ecosystems
and man come to somewhat tenuous terms. Increased numbers of
people attempting to live off the productivity of native
vegetation or cultivated crops in high mountains have often
led to a forgetting of lessons learned and in many cases to a
breaking of the terms, and resultant degradation of the
ecosystems and social systems.

If man and his hooved animal introductions damage the
native mountain vegetation beyond its carrying capacity, accelerated erosion is sure to follow. This is particularly
evident in those high mountain ecosystems on islands where
there were no hooved mammals in the natural ecosystems; New
Zealand is an example. But it is also occurring in continental
mountains where there were native mammalian herbivores in the
system. Severe erosion in the Andes is testimony to increases
in the number of people and animals taking the original
ecosystems beyond their carrying capacities. These carrying
capacities are based on primary productivities upon which
both aboveground and belowground plant growth depend. New
Zealand and the Andes are but examples; the same thing is

taking place in most high mountain regions on all continents.

Characteristics Held in Common by High Mountain Ecosystems

At this point, it is appropriate to list what all or most high mountain ecosystems have in common.

1. Physical Components
   a. All have relatively low daytime air temperatures.
   b. If not in the tropics, all have relatively high precipitation amounts as compared to lowlands. Much of this falls as snow.
   c. Atmospheric pressure is lower with increased elevation.
   d. Partial pressures of the metabolic gases O_2 and CO_2 are lower with increased elevation.
   e. Most have relatively high solar and terrestrial radiation flux rates relative to the regional lowlands.
   f. Wind speeds are higher than in adjacent lowlands.
   g. Most have long-lasting snowdrifts and/or ice.
   h. Plant growing seasons vary from short to long, but number of degree-days above $0°C$ is low compared with adjacent lowlands.
   i. Steep, unstable slopes are the rule.
   j. Soils are mostly shallow and susceptible to erosion.
   k. Solifluction and soil frost activity are common.
   l. All have a high degree of habitat diversity, or environmental patchiness, because of the interaction of elevational, mesotopographic, and microtopographic gradients.

2. Biological and Evolutionary Components
   a. There are limited numbers of options for:
      a.1 - plant life forms
      a.2 - plant metabolic processes
   b. Endemism is high in plant species and certain animal taxa.
   c. There can be relatively rapid rates of adaptive radiative evolution in genetically pre-adapted taxa.
   d. All high mountains have biological communities which are unique to relatively small and specific areas.
   e. All natural high mountain ecosystems evolved in the absence of man, and many in the absence of hooved mammals.

3. Aesthetic Components
   a. High mountains have certain "artistic" values.
   b. These values lead to use by tourism and recreation which in themselves are economic components.

4. Economic Components
   a. All mountains yield water as runoff.
   b. Most mountains yield some forage and/or timber.
   c. Most mountains contain useable minerals.

5. Scientific Components or Opportunities
   a. The steep environmental gradients on mountainsides are ideal sites to study absolute tolerance ranges of certain plant and animal taxa.
   b. Such tolerance range studies could involve using mountains to determine temperature and aridity limits of economic plants including hybrids or ecotypes which might extend their economic ranges.
   c. Mountain ecosystems should be studied because they are different from all other ecosystems, even those of the Arctic. They are unique and, as with islands, their boundaries are discrete.
   d. Because daytime air temperatures are about the same in all high mountains above timberline, they can be used as "outdoor phytotrons" for certain kinds of plants along latitudinal, moisture, photoperiodic, and soil gradients.
   e. High mountains can be used as "islands" in experimental biogeography.
   f. The study of elevational migrations of known genotypes and ecotypes could provide sensitive indices to changes in global and regional environments.

Research Priorities on High Mountain Ecosystems

In spite of at least a century of ecological work in high mountains, alpine ecosystems are really not known either in structure or operation. Moreover, modes of maintenance of these systems in good order are not known. We have only some general rules of thumb which may work for a while on some mountains but not on others. The following list of research priorities on high mountain ecosystems is not complete and the ordering is of no great significance--but they would help us to understand and live with these unique systems. Here are some of the principal needs:

1. Detailed mapping of mountain ecosystems.
2. Censusing of plant and animal species in high mountains over the earth.

3. Mapping and censusing of permanent sample areas to measure changes in biota through time in both peopled and unpeopled mountains.
4. Information on primary and secondary productivity, biomass, and food webs in alpine ecosystems of all types. The dearth of such knowledge is all too apparent.
5. Intensive studies on plant adaptations, both physiological and morphological, to high altitude environments.
6. Vegetational structure data along mesotopographic gradients in relation to water availability, temperature effects, length of growing season, and soil effects. Data of this type can be used to compare mountain ecosystems along a somewhat standard gradient.
7. Controlled *whole* ecosystem operational measurements before and after perturbation. These should be on the scale of the Hubbard Brook work in New England (45, 46, 47) and incorporate the best techniques of that work with the best ideas emerging from IBP studies of arctic tundras and alpine ecosystems.
8. Ecological studies of high mountain agricultural systems.
9. Studies on the impact of people on natural alpine ecosystems.
10. Intensive research on the relationship between people, vegetation, soil, and water yield in light of the spread of accelerated erosion in high mountains.
11. The setting aside of some large mountain ecosystem reserves, fully protected, for long-term scientific baseline studies. These would be the monitored controls not only for comparison with impacted mountains but as sensitive detectors of biospheric changes.
12. The establishment of a network of high mountain research stations for ecological work; the number of these available over the earth now is but an inadequate handful.
13. A start should be made on systems analysis and systems modelling of *whole* mountain ranges.

Concluding Statement

Since high mountain ecosystems and their biota are particularly vulnerable to the presence of people, one must conclude from observations to date that there is a high degree of incompatibility between use by people and the maintenance of the integrity of these systems. The original ecosystems have already been degraded by such activity in most mountains; this degradation is accelerating and spreading.

Therefore, it is imperative that a serious attempt be made to educate the general public to the facts that high mountain ecosystems are inherently unstable, and that they are particularly sensitive to the presence of large numbers of people and their activities. The uses of these ecosystems must be carefully managed for renewable resources; water yield, plant and animal production, recreational and aesthetic values. For education, and management, we must have solid ecological facts as well as theory.

Acknowledgments

I thank the General Ecology Program, Ecosystems Analysis Program, and Division of Polar Programs of the National Science Foundation for grants which enabled me to study arctic and alpine ecosystems over a period of years. I appreciate the help of many colleagues and students in this work. Particular appreciation is due to Shirley M. Billings and Patrick J. Webber for encouragement and understanding in the preparation of this paper.

References

1. Fig. 1 from H. Walter. *The Vegetation of the Earth*. Springer-Verlag, New York: 237 pp. (1973). Reproduced by permission.

2. F. Vuilleumier, Insular biogeography in continental regions. I. The northern Andes of South America. *Amer. Natur.*, 104: 373-388 (1970).

3. B.S. Simpson, Pleistocene changes in the flora of the high tropical Andes. *Paleobiology*, 1: 273-294 (1975).

4. J.H. Brown, Mammals on mountaintops: nonequilibrium insular biogeography. *Amer. Natur.*, 105: 467-478 (1971).

5. W.D. Billings, Alpine phytogeography across the Great Basin. *In* K.T. Harper and J.L. Reveal (eds.), *Intermountain Biogeography: A Symposium, Great Basin Naturalist Memo. 2*, in press (1977).

6. B.F. Chabot and W.D. Billings, Origins and ecology of the Sierran alpine flora and vegetation. *Ecol. Monogr.*, 42: 163-199 (1972).

7. S. Carlquist, *Island Life*. Natural History Press, New York: 451 pp. (1965).

8. D. Amadon, The Hawaiian honeycreepers (Aves, Drepaniidae). *Amer. Mus. Natur. Hist. Bull.*, 95(4): 155-262 (1950).

9. G. Ruhle, *A Guide to the Crater Area of Haleakala National Park.* Nat. Park Serv., Hawaii Nat. Hist. Assoc.: 78 pp. (1975).

10. G.C. Munro, *Birds of Hawaii.* Tongg Publ. Co., Honolulu: 189 pp. plus 20 plates (1944).

11. J.A. Teeri and L.G. Stowe, Climatic patterns and the distribution of C_4 grasses in North America. *Oecologia,* 23: 1-12 (1976).

12. T. Harrison, personal communication (Aug. 23, 1977).

13. H. Reisigl and H. Pitschmann, Obere Grenzen von Flora und Vegetation in der Nivalstufe der Zentralen Ötztaler Alpen (Tirol). *Vegetatio,* 8: 93-129 (1958).

14. S.-W. Breckle, Notes on alpine and nival flora of the Hindu Kush, East Afghanistan. *Bot. Notiser,* 127: 278-284 (1974).

15. W.D. Billings, Adaptations and origins of alpine plants. *Arct. Alp. Res.,* 6(2): 129-142.

16. W.D. Billings, Arctic and alpine vegetations: similarities, differences, and susceptibility to disturbance. *BioScience,* 23(12): 697-704 (1973).

17. D. Scott and W.D. Billings, Effects of environmental factors on standing crop and productivity of an alpine tundra. *Ecol. Monogr.,* 34: 243-270 (1964).

18. W.D. Billings and H.A. Mooney, The ecology of arctic and alpine plants. *Biol. Rev.,* 43: 481-530 (1968).

19. L.C. Bliss, Arctic and alpine plant life cycles. *Ann. Rev. Ecol. Syst.,* 2: 405-438 (1971).

20. W.D. Billings, Arctic and alpine vegetation: plant adaptations to cold summer climates. *In* J.D. Ives and R.G. Barry (eds.), *Arctic and Alpine Environments.* Methuen, London, pp. 403-443 (1974).

21. J. Major and D.W. Taylor, Alpine. *In* M.G. Barbour and J. Major (eds.), *Terrestrial Vegetation of California.* Wiley-Interscience, New York: 601 (1977).

22. P.J. Webber, Tundra primary productivity. *In* J.D. Ives and R.G. Barry (eds.), *Arctic and Alpine Environments*. Methuen London: 445-473 (1974).

23. M. Vetter, personal communication (July 12, 1977).

24. W.D. Billings, P.J. Godfrey, B.F. Chabot, and D.P. Bourque, Metabolic acclimation to temperature in arctic and alpine ecotypes of *Oxyria digyna*. *Arct. Alp. Res.*, 3(4): 277-289 (1971).

25. L.L. Tieszen and J.A. Helgager, Genetic and physiological adaptation in the Hill Reaction of *Deschampsia caespitosa*. *Nature*, 219: 1066-1067 (1968).

26. E. Cartellieri, Über Transpiration und Kohlensaureassimilation an einem hochalpinen Standort. *Sber. Akad. Wiss. Wien Math.-Nat. Klasse*, 149: 95-143 (1940).

27. W.D. Billings, E.E.C. Clebsch, and H.A. Mooney, Photosynthesis and respiration rates of Rocky Mountain alpine plants under field conditions. *Amer. Mid. Natur.*, 75(1): 34-44 (1966).

28. H.A. Mooney and W.D. Billings, The annual carbohydrate cycle of alpine plants as related to growth. *Amer. J. Bot.*, 47: 594-598 (1960).

29. E.B. Hadley and L.C. Bliss, Energy relationships of alpine plants on Mt. Washington, New Hampshire. *Ecol. Monogr.*, 34: 331-357 (1964).

30. A. Cernusca and W. Moser, Die Automatische Registrierung Produktions-Analytischer Mebdaten bei Freilandversuchen auf Lochstreifen. *Photosynthetica*, 3(1): 21-27 (1969).

31. W. Moser, Ökophysiologische Untersuchungen an Nivalpflanzen. *Mittl. Ostalp.-din. Ges. f. Vegetkde.*, 11: 121-134 (1970).

32. W. Moser and C. Licht, Temperatur und Photosynthese an der Station "Hoher Nebelkogel" (3184 m). *In* H. Ellenberg (ed.), *Ökosystemforschung*. Springer-Verlag, Berlin: 203-223 (1973).

33. L.C. Bliss, Plant productivity in alpine microenvironments on Mt. Washington, New Hampshire. *Ecol. Monogr.*, 36: 125-155 (1966).

34. F.E. Wielgolaski, Primary productivity of alpine meadow communities. *In* F.E. Wielgolaski (ed.), *Fennoscandian*

Tundra Ecosystems, Part 1, Plants and Microorganisms.
Ecol. Studies 16, Springer-Verlag, Berlin: 121-128 (1975).

35. D.W. Gade, *Plants, Man, and the Land in the Vilcanota Valley of Peru.* Dr. W. Junk, The Hague: 240 pp. (1975).

36. R.S. Hoffman, Terrestrial vertebrates. *In* J.D. Ives and R.G. Barry (eds.), *Arctic and Alpine Environments.* Methuen, London: 475-568 (1974).

37. L.W. Swan and A.E. Leviton, The herpetology of Nepal: a history, check list, and zoogeographical analysis of the herpetofauna. *Proc. Calif. Acad. Sci.*, 32: 103-147 (1962).

38. D.L. Pattie and N.A.M. Verbeek, Alpine birds of the Beartooth Mountains. *Condor*, 67: 167-176 (1966).

39. D.L. Pattie and N.A.M. Verbeek, Alpine mammals of the Beartooth Mountains. *Northwest Sci.*, 41: 110-117 (1967).

40. M.S. Mani, *Introduction to High Altitude Entomology. Insect Life Above the Timber-line in North-west Himalaya.* Methuen, London: 302 pp. (1962).

41. F.E. Wielgolaski (ed.), *Fennoscandian Tundra Ecosystems, Part 1, Plants and Microorganisms.* Ecol. Studies 16, Springer-Verlag, Berlin: 366 pp. (1975).

42. F.E. Wielgolaski (ed.), *Fennoscandian Tundra Ecosystems, Part 2, Animals and Systems Analysis.* Ecol. Studies 17, Springer-Verlag, Berlin: 337 pp. (1975).

43. W.D. Billings, Vegetational pattern near alpine timberline as affected by fire-snowdrift interactions. *Vegetatio*, 19: 192-207 (1969).

44. V.C. LaMarche and H.A. Mooney, Altithermal timberline advance in Western United States. *Nature*, 213(5080): 980-982 (1967).

45. G.E. Likens, *et al.*, *Biogeochemistry of a Forest Ecosystem.* Springer-Verlag, New York: 146 pp. (1977).

46. F.H. Bormann, *et al.*, The Hubbard Brook ecosystem study: composition and dynamics of the tree stratum. *Ecol. Monogr.*, 40: 373-388 (1970).

47. F.H. Bormann, *et al.*, The effect of deforestation on ecosystem export and the steady-state condition at Hubbard Brook. *Ecol. Monogr.*, 44: 255-277 (1974).

7

High Altitude Physiology

Robert F. Grover

If you were sitting on the shore of a Pacific Island, you would not only be at the surface of the ocean of water but you would also be at the bottom of the ocean of air which envelopes the earth. This air has mass, it has weight, and it exerts pressure (the atmospheric pressure), which at sea level is 15 pounds per square inch, or sufficient to support a column of mercury 760 mm high. As we ascend through the atmosphere, there is progressively less air above us and more below us, so the atmospheric pressure decreases. That relationship is shown in Figure 1 (1). Since the atmosphere is approximately one-fifth oxygen, then the partial pressure of oxygen would decline in a similar fashion.

The physiological effects of high altitude are classically attributed to this reduction of inspired oxygen pressure (atmospheric hypoxia) and the resultant decreases in oxygen pressures within the body. When you go to high altitude, there are only two possible ways in which you can avoid the effects of atmospheric hypoxia. One is to be inside a large high-pressure chamber such as a jet aircraft, so that at an external altitude of 12,000 m you remain in complete comfort because the total pressure of air within the cabin is equivalent to a much lower altitude. The other alternative is to supplement the air you are breathing with oxygen if you are willing to carry an oxygen tank on your back and wear a mask, and in this way you can avoid the reduction in inspired oxygen pressure. This technique is employed when climbing the high peaks of the Himalaya. Both of these techniques are useful for short periods of time, but the vast majority of people who go to high altitudes to live accept the atmospheric hypoxia and must adapt to it.

An impressively large segment of the human population is exposed to high altitude. Weiner (2) has estimated that

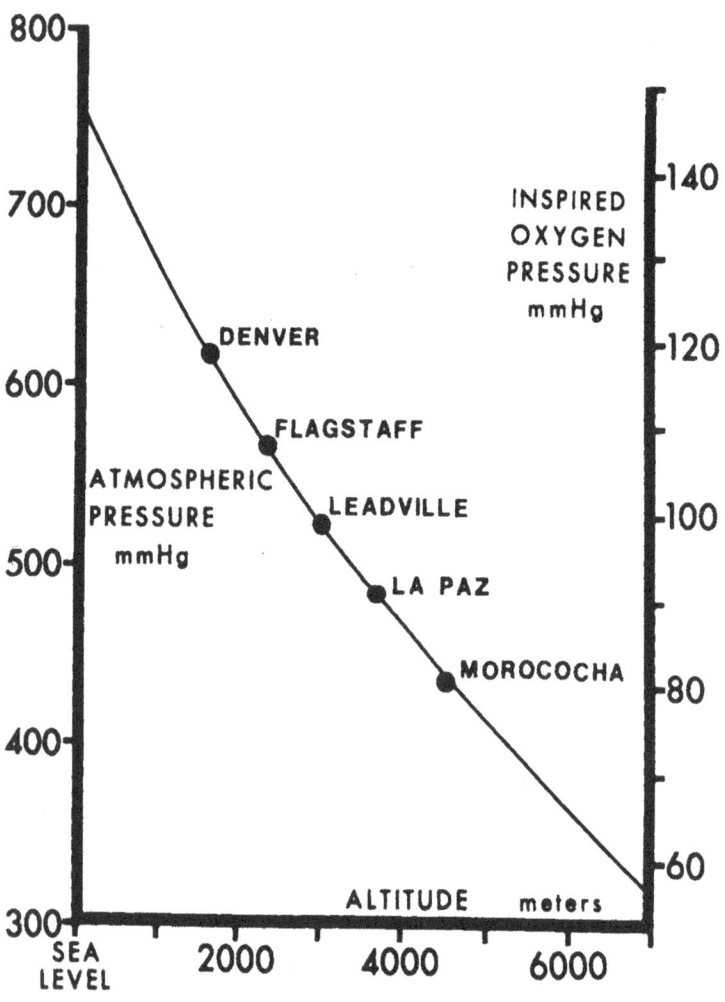

Figure 1.   Atmospheric and oxygen pressure as a function of
altitude; five high altitude locations are plotted
on the curve of this relationship.

approximately 12% of the world's population lives in mountainous areas, and DeJong (3) estimates that some 25,000,000
people live at altitudes higher than 3000 m. That estimate
was made ten years ago, and probably the number is much
higher now. Probably the largest municipality at really
high altitude is La Paz, Bolivia, at about 3500 m. One of
the reasons why such large populations can exist in the
Andes is that these high regions are located close to the
equator where the climate is quite favorable.

Here in Colorado we have the city of Leadville, approximately two miles high (3100 m). This is a community of
8000 people in a high mountain valley roughly 170 km west of
Denver in a spectacular setting of high peaks surrounding
the city. Leadville was established as a mining town roughly
100 years ago, and so the natives have been there no more
than four generations. This is in contrast to the population
in the Andes that has been at high altitudes since roughly
7500 B.C. (4). Even they are relatively new to high altitude
compared to the inhabitants of the high Tibetan plateau;
they probably go back 200,000 years, much longer than any
people in the Andes. High altitude populations are also
located in the Himalaya, eastern Africa, and the Caucasus of
southern USSR. In addition to these permanent residents, we
have vast numbers of visitors to high altitude. In Colorado,
literally millions of people come for various recreational
purposes such as climbing, fishing, hunting, and skiing.
Pikes Peak has a highway to the summit at 4300 m, and somewhat
less well known but equally high is Mt. Evans, immediately
west of Denver, also 4300 m, again with a highway to the
summit. On an average summer day 3600 people visit the
summit of Pikes Peak! That means that in an average summer
season, a quarter of a million people visit these two mountain
areas alone. The ski areas are also at considerable altitude,
e.g., at Breckenridge the top of the mountain is 3600 m, and
the town itself is at 2900 m. Hence we have people from
near sea level coming to respectable altitudes and engaging
in fairly strenuous exertion such as skiing. Since we are
dealing with a rather large segment of the human population
that is exposed to high altitudes for shorter or longer
periods of time, the consequences are worthy of consideration.

Physiologically the most significant impact of high
altitude is the reduction in man's capacity to perform
muscular work. In sustained exercise of longer than a few
minutes duration, the working muscles must have a continuous
supply of oxygen. Exercise physiologists quantitate aerobic
working capacity as the maximum amount of oxygen that an
individual can consume in a minute when the exertion is of
only 3 to 5 minutes duration, usually on a treadmill. Many

investigators have reported that this maximum oxygen uptake
is reduced at high altitude. For example, we found (5) a
reduction of 20% in a group of athletes taken to 3100 m
altitude for three weeks. Buskirk (6) collected such
published data and came up with a general relationship for
the decline in maximum oxygen uptake with increasing altitude
(Figure 2). There are two significant features to this
relationship. From sea level to roughly Denver's altitude
(1609 m) there is no measurable reduction in maximum oxygen
uptake, so Denver seems to be at the threshold for this
phenomenon. Above 1500 m, there is a progressive reduction
in maximum oxygen uptake at the rate of 10% per 1000 m.
Thus, for a laborer working on the second bore of the
Eisenhower Tunnel under Loveland Pass at 3400 m, his capacity
for oxygen uptake will be only 80% of what it was at sea
level.

Now what does this mean in terms of a person's ability
to work at high altitude? If you are an athlete running at
the limits of your capacity against the clock, this reduced
aerobic capacity is going to impair your performance, i.e.,
at the higher altitude it will take longer to cover a given
distance. That is the effect on the person who is doing
something for only five minutes, but what does reduced
aerobic capacity mean to the individual who is not making an
all-out effort, but rather who is performing a normal day's
physical activity at high altitude? The answer to that is
not really known precisely. Astrand and Rodahl (7) found
that if you measure maximum oxygen uptake in a brief 5-
minute period of exertion and then measure what that same
individual can do if he were required to work for a matter
of hours, the longer he works, the lower the percent of
maximum he is able to sustain (Figure 3). At the very best,
in an eight-hour day, a person could work at roughly 50% of
his capacity. However, I think that for most people it
would be somewhat less, about 25% of capacity for eight
hours. It is that kind of relationship that applies to a
man who is performing more usual types of physical activity.
To illustrate, suppose the individual is a carpenter, or a
brick layer, or a miner working at the Climax Molybdenum
mine above Leadville at about 3400 m, where during mining
operations the individuals go to even higher altitudes.
What is the effect of this kind of altitude on the laborer?
We have to make some estimates based on the above information.
At 3400 m, if you are limited in the number of people that
you can put on a job, then at this altitude that work force
is going to take 4 days to do what they would normally do in
3 days at sea level, or 40 days instead of 30 days; or if
you are digging a tunnel under the Continental Divide, 4 years
instead of 3 years. The other alternative is that, if you

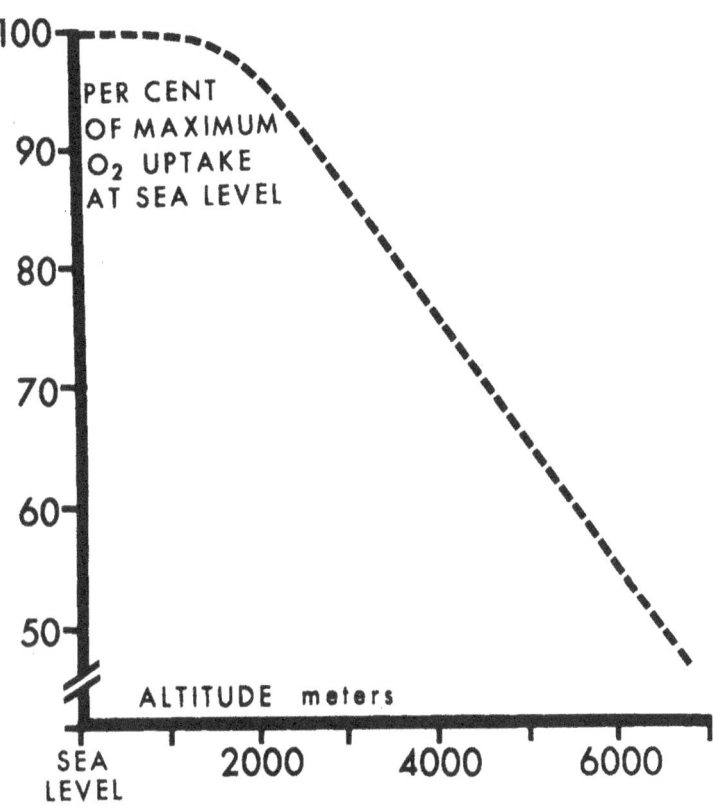

Figure 2.   The decline in maximum oxygen uptake as a function of altitude (adapted from Buskirk, 6).

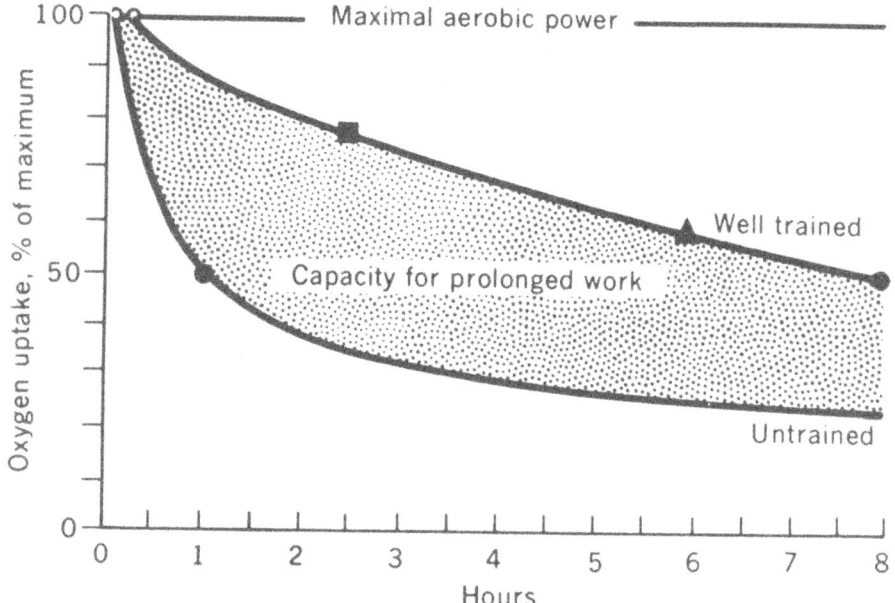

Figure 3.   A comparison of the ability of well-trained and
untrained workers to take up oxygen after varying
periods of work.

are pressed for time, you must add to the number of people
you put on the job. You will then need 4 people to do the
job done by 3 at sea level, or 40 instead of 30 at this
altitude. Either way it takes more man-hours to do a given
task at high altitude than at sea level and that costs more
money. The economic impact of altitude on construction
projects is great. These are only rough approximations, and
it is important that better measurements be made under these
types of situations.

Having considered the implications of a reduced working
capacity at high altitude, we will now examine the physiological alterations which account for this reduction in your
capacity to consume oxygen and do aerobic work. We can say
at the beginning that this decreased capacity to take up
oxygen is related to a decreased ability of the body to
deliver oxygen to the working muscles. Oxygen delivery
involves three major processes. First is pulmonary ventilation, which gets the air and oxygen from outside to the
inside of the body. Once the oxygen is in the lungs, it
then diffuses into the bloodstream where it is bound by
hemoglobin. Then the oxygenated blood is pumped by the
heart through the vascular system to the muscles. Briefly
we will look at these three components of the oxygen transport
system in an effort to identify the weak link in the chain.

Let us consider the question of the ability of the
lungs to move air. We all know that when you go to high
altitude ventilation increases. Is that the limiting factor?
To examine this, individuals worked on a treadmill at a
constant speed, and the grade was increased progressively
until the person became exhausted. As the work load increased, the ventilation and the oxygen uptake increased
until we reached the person's maximum. Then further increase
in the work load resulted in no further increase in oxygen
uptake; however, ventilation continued to climb, but in
spite of this added ventilation there was no further increase
in oxygen uptake (5). This says that at moderate altitudes,
certainly like 3100 m, ventilation is not the limiting
factor.

Now the next step in the chain is the transfer of
oxygen from air to blood. This takes place at the terminal
air sack in the lung, the alveolus. Here the red blood
cells sliding through the capillary come very close to the
air in the alveolus. The hemoglobin, the iron pigment in
the red cell, combines with this oxygen; the amount of
oxygen bound is related to the oxygen pressure by the
relationship in Figure 4. When the pressure is high, as it
is at sea level, the hemoglobin is almost fully saturated at

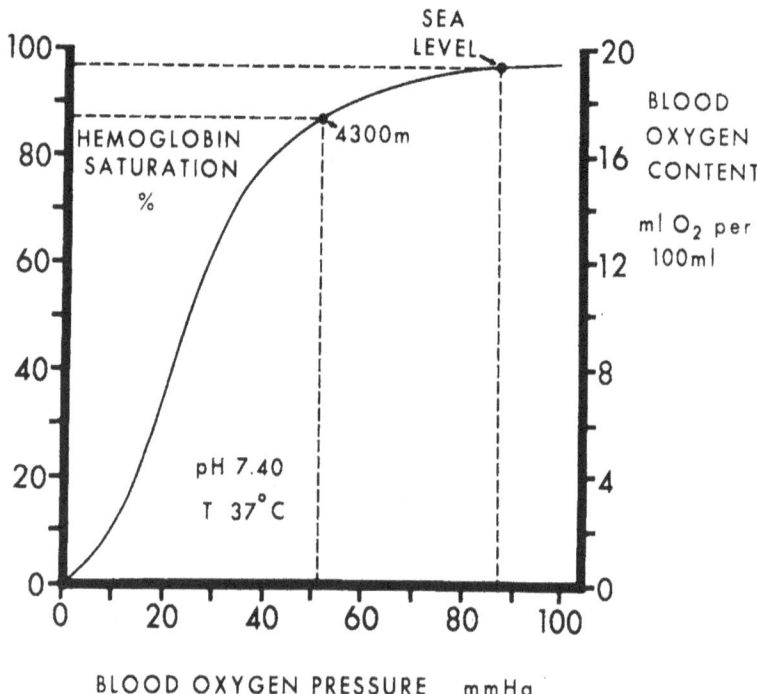

Figure 4.   Oxygenation of arterial blood at sea level and
high altitude.

98%. With ascent to altitude, as the pressure of oxygen declines, at first there is only a very small decrease in saturation; it is only when you get to very high altitudes that saturation begins to fall markedly. At 4300 m, the blood is about 85% saturated; and in the Leadville situation at 3100 m, the saturation is actually about 90%. Thus, in spite of the fall in oxygen pressure, the quantity of oxygen bound to the hemoglobin remains quite high, a very well engineered system. Hence, blood oxygenation is not normally a limiting factor at altitudes up to 4000 m.

That brings us down to the circulation of the oxygenated blood by the heart, the so-called "cardiac output." What happens to this quantity when man goes to high altitude? Cardiac output is directly related to oxygen uptake which is increased as exercise load in increased (Figure 5). You work harder, you need more oxygen, you circulate more blood. In individuals taken to 3100 m, after 10 days there is a nearly 20% reduction in the amount of blood pumped by the heart per minute in healthy young individuals (8). This then, is the weak link in the system. You cannot do as much work because you cannot deliver as much oxygen because your heart does not pump as much blood. You get plenty of oxygen in the blood; you just cannot circulate enough of it to do as much work as you did at sea level. The fact that your heart does not pump as much does not reflect any damage to the heart or even an inadequate supply of oxygen to the heart muscle. Rather, this is probably a consequence of circulatory changes which reduce the volume of blood returned to the heart for pumping (9). This is a totally innocuous response as far as the heart is concerned. It is completely reversible when you go back to low altitude. But as long as you stay at high altitude, the quantity of blood pumped by your heart remains less than it was at sea level.

Up to this point we have considered what happens to the man from sea level when he adapts to high altitude. What can we say about people who have been at high altitudes for many generations, as in the high Andes? They have a remarkable capacity for exercise, and after doing a normal day's work, in the afternoon they go out and play soccer just for recreation at 4300 m! As a group they have very high capacities for exercise, comparable to that of special athletic-type individuals we find at low altitude. The other unusual thing about them is that changing altitude has much less influence on their capacity for work. This is primarily because changing altitude in these people does not produce the alterations in cardiac output observed in the men from sea level going to high altitudes (10). Just why that is, we do not know; but that seems to be one physio-

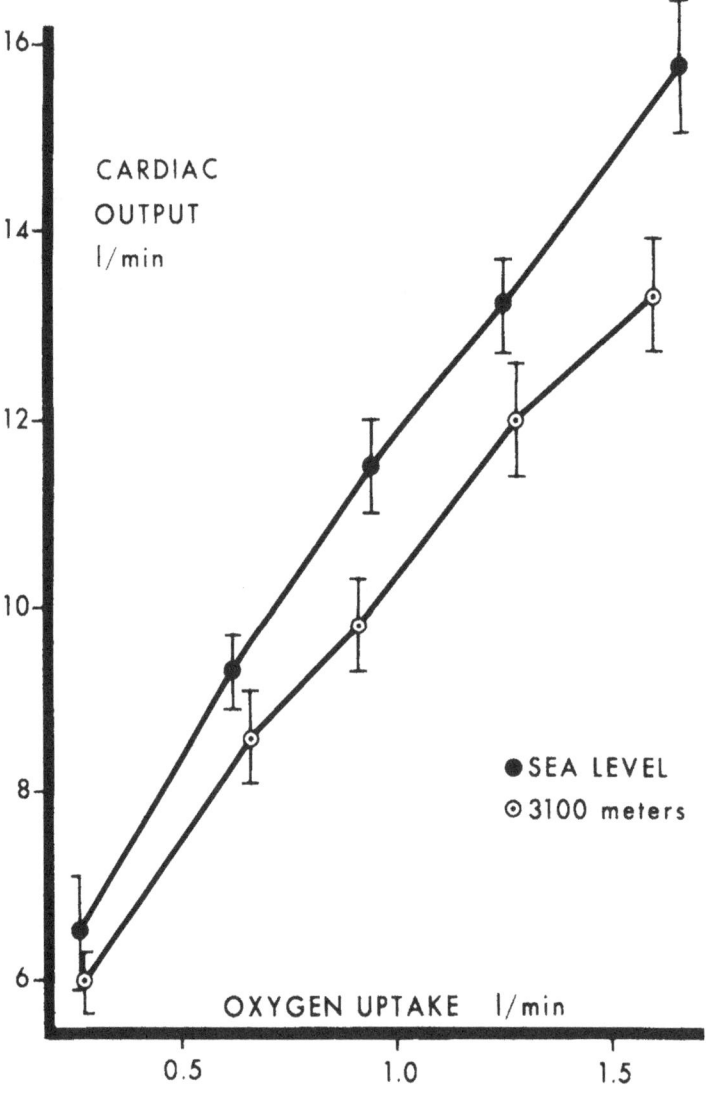

Figure 5.   Circulation of oxygenated blood by the heart or cardiac output as a function of oxygen uptake at sea level and 3100 meters above sea level.

logically important distinguishing feature that makes these people so remarkable.

Even the high altitude native is not immune from altitude problems. One of the things that happens to some men in the Andes (and some men in Leadville) is that they lose their acclimatization and develop a condition known as chronic mountain sickness. This is the physiological consequence of not getting enough oxygen into the blood. That occurs primarily during sleep when these people under-breathe markedly (11). The body responds to this oxygen deficiency by producing an excessive quantity of red blood cells (polycythemia). The symptoms of chronic mountain sickness are sluggishness, headache, generally feeling poorly, and impaired mental function. You can relieve those symptoms temporarily by removing some of the excessively thick blood, but soon the body produces more red blood cells to replace what was removed, so that is not a cure. Obviously you can take the individual to low altitude, and the condition then clears up spontaneously. As an alternative form of treatment, we are currently working with a substance which stimulates breathing (medroxyprogeserone acetate), particularly during sleep, and are having a rather remarkable degree of success. This condition is apparently completely reversible if you simply get people to breathe well during sleep (11).

In conclusion, I have presented some of the physiological consequences of going to high altitude. Of major importance is the reduction in physical work output of individuals, requiring them to do all things more slowly. That is not necessarily undesirable, and I think most of us are willing to accept that as a small price to pay for the aesthetic values of mountain life.

References

1. D.B. Dill and D.S. Evans, Report barometric pressure! *J. Appl. Physiol.*, 29: 914 (1970).

2. J.S. Weiner. In: *Human Biology* (G.A. Harrison, J.S. Weiner, N.A. Barnicot, and J.M. Tanner, Eds.). Clarendon Press, Oxford (1964).

3. G.F. DeJong, The demography of high-altitude populations. (Report to W.H.O./P.A.O.H./I.B.P. Meeting of Investigators on Population Biology of Altitude. Pan-American Health Organization, Washington, D.C.) (1968)

4. T.F. Lynch, Quishqui Puncu: a preceramic site in highland Peru. *Science*, 158: 780 (1967).

5. J.T. Reeves, R.F. Grover, and J.E. Cohn, Regulation of ventilation during exercise at 10,200 ft. in athletes born at low altitude. *J. Appl. Physiol.*, 22: 546 (1967).

6. E.R. Buskirk, Decrease in physical working capacity at high altitude. In: *Biomedicine Problems of High Terrestrial Elevations* (A.H. Hegnauer, Ed.). U.S. Army Research Institute of Environmental Medicine, Natick, Mass. (1969).

7. P.O. Astrand and K. Rodahl, *Textbook of Work Physiology*, McGraw-Hill, New York: 292 pp. (1970).

8. J.K. Alexander, L.H. Hartley, M. Modelski, and R.F. Grover, Reduction of stroke volume during exercise in man following ascent to 3,100 m altitude. *J. Appl. Physiol.*, 23: 849 (1967).

9. R.F. Grover, J.T. Reeves, J.T. Maher, R.E. McCullough, J.C. Cruz, J.C. Denniston, and A. Cymerman, Maintained stroke volume but impaired arterial oxygenation in man at high altitude with supplemental CO_2. *Circ. Res.*, 38: 391 (1976).

10. J.A. Vogel, L.H. Hartley, and J.C. Cruz, Cardiac output during exercise in altitude natives at sea level and high altitude. *J. Appl. Physiol.*, 36: 173 (1974).

11. M. Kryger, R.E. McCullough, R. Doekel, D. Collins, J.V. Weil, and R.F. Grover, Excessive polycythemia of high altitude: the role of lung disease and decreased ventilatory drive. *J. Appl. Physiol.* (in press)

8

Effects of Change on High Mountain Human Adaptive Patterns

R. Brooke Thomas

Introduction

In the course of the past decade, it has become increasingly apparent that many human and environmental systems in mountainous areas are undergoing an unprecedented rate of deterioration which may well be irreversible for the foreseeable future. The consequences of this do not just remain in the highlands, but effect the lands below which receive downslope flows of dislodged soil, substrate, and people (1).

The seriousness of the problem has been most clearly summarized by Eckholm (2) in a recent review on the condition of the world's mountain environments. He states:

> On the basis of already available knowledge, it is no exaggeration to suggest that many mountain regions could pass the point of no return within the next two or three decades. They could become locked in a downward spiral from which there is no escape, a chain of ecological reactions that will permanently reduce their capacity to support human life (p. 769).

While such deterioration has become clear to scientists and governmental administrators alike, there is little consensus on how long these conditions have persisted, or what general factors underlie them. Lowlanders are just becoming aware of the magnitude of the problem. In the past mountainous regions have often been regarded as peripheral (3) except when they have provided specific resources to the lowlands. As a consequence, no systematic body of information exists on which to draw. When information is available, it is largely descriptive in nature; this is an inadequate data base for understanding the problem at hand or initiating solutions.

There is irony here. Lowlanders who have acquired such
a meager knowledge of the operation of mountain human/environ-
mental systems, and who may in part be responsible for creating
the deteriorating conditions, now see it as their responsibility
to reverse this trend. Whether the motivations underlying
their recent concern are genuinely altruistic or more opportunistic does not particularly matter. The point is that policy
oriented at ameliorating highland conditions will, by and
large, be directed by lowland institutions. Thus one can
only assume that lowland priorities and interests will not be
violated.

This reality aside, it matters very much the human and
environmental consequences of such policies and the programs
designed to carry them out. Unfortunately with so little
known about 1) the complex environmental conditions existing
in mountainous regions, 2) the biocultural responses which
have allowed humans to adjust to these conditions, and 3) the
scale and rate with which deterioration is taking place, it
is difficult to assess presently the consequences of change.

This is the dilemma of administrators of planned change.
On one hand they are being urged to take action; on the other
hand they frequently do not have access to an adequate data
base or, more important, a systematic conceptual or analytical
framework upon which to base decisions. As a result there is
little assurance that programs designed to improve specific
conditions will not ultimately worsen the overall situation.

A thorough collection of data concerning structure,
function, and process in mountain human/environmental systems
is critical. Despite the desirability of this as a long-term
approach, the urgency of the situation suggests that policy
programs will be pushed ahead based on information presently
available. If this is to be the case, it is appropriate to
establish a framework which can utilize existing knowledge
and provide a basis for anticipating some of the more obvious
and disruptive consequences of change.

Briefly, this framework is developed around the concept
that living systems have limits in their ability to adjust to
environmental conditions. Thus the ability of an organism or
population to respond to conditions in such a way as to
maintain critical components or variables within their limits
can broadly be referred to as "adaptation." The concept of
adaptation fundamentally is the process whereby adaptive
units persist through time (4, 5), persistence being measured
by the continued presence of identifiable characteristics of
the unit. This is a relevant framework for understanding
mountain systems and the types of change which could exceed

their adaptive capabilities.

Such a framework identifies and provides adaptive generalizations about critical interacting variables in highland/
environmental systems. Attention in this paper will be upon
the human subsystem and those environmental conditions that
bear directly upon its adaptiveness or well-being. This is
not to say that a knowledge of ecological structure, function,
and process is not important, but rather that human populations
and their interactions with the environment are of principal
concern. Given the environmental complexity of highland
regions, and the time and effort it would take to gather
basic ecological data, it seems more prudent to focus upon
humans, their resources, and the environmental factors that
directly impinge on their ability to produce these resources.

While environmental scientists have paid considerable
attention to what humans are doing wrong to their physical
and biotic environment, the present paper will take an
opposite approach, based on the assumption that human populations, having long-term exposure to a region, are aware of
its environmental problems and opportunities and have made
appropriate adjustments. This, of course, is not to say that
everything populations do is adaptive, or that recent changes
in environmental conditions have not altered the effectiveness
of some responses. Thus, I will attempt to identify what
humans do right in adjusting to the conditions of mountainous
regions, and how certain changes are capable of disrupting
these adaptive responses.

Humans living in heterogeneous and unpredictable mountainous environments serve as particularly rich examples
since they provide insights into how an adaptive system based
on phenotypic plasticity adjusts to extreme spatial and
temporal diversity. Ecological theory suggests that organisms
exposed to such conditions would display highly flexible
adaptive responses. Furthermore, phenotypic plasticity
allows greater flexibility in the face of rapidly changing
conditions than does genetic change. Apparently the recent
changes occurring in highland human groups have either
created conditions which exceed the capacity of their presumably flexible biobehavioral adaptive responses or have eroded
their effectiveness. It is of considerable interest (though
unfortunate for the participants in the system) to study the
response limitations of an adaptive system in adjusting to
rapid change. Knowing what such limitations are, in turn,
contributes to a conceptual framework which has applied
value.

Orientation

It has been estimated that mountain areas lying above 1000 m elevation make up a quarter of the earth's surface and contain 10% of the human population. While this is approximately 350 million people, another 1400 million inhabitants live in adjacent lowland areas and are highly dependent upon the condition of the human/environmental systems above them (2). High mountains are designated here as regions lying above 3000 m in elevation. The human population residing at these altitudes is between 20 and 25 million inhabitants (6). Since environmental conditions and human responses characterizing mountainous regions are accentuated in the high mountains, and because human groups residing there are isolated from lowland systems, high mountain regions will be the primary concern of this paper.

Principal high mountain systems of the world include the Andes, Rocky Mountains, Alps, Caucasus, Pamir, Tien Shan, Karakorams, and Himalaya. In addition a number of elevated upland areas exist, such as parts of East Africa and the Middle East, which can be classified as high altitude regions (7). Since the zones of permanent human habitation are most elevated in the tropical and subtropical high mountains (subsequently referred to as "tropical high mountains") these will receive major attention. Similarities in conditions and human responses exist at lower elevations in the mid-latitude mountains, and this material will be utilized when it appears consistent with that available for the tropical high mountains.

Selection of high mountain systems for this paper has depended upon the adequacy of the data base. While the paucity of information on any given high tropical mountain range is clearly evident, a number of excellent publications summarizing and comparing these data have recently appeared and generalizations from these will be heavily relied upon. These attempt to synthesize knowledge of the environment (1, 2, 8-13), human biology (7, 14-17), and human social systems (3, 18-20). These publications reflect not only a growing interest in problems in high mountain regions, but also an increasing awareness of the dynamic interactions operating in highland human/environmental systems. However, since these are based primarily upon information from either the Andes or Himalaya, there is bias in the present analysis. Because considerable information is available on human ecology in the Alps, these data, when appropriate, will also be used.

The Andes and Himalaya have substantial human populations residing permanently at high altitude, and they both are traditionally non-Western systems which are being exposed to

Western ideas of modernization, development, and progress. Comparisons between the two ranges are instructive since the Andes has had a more excessive recent history of exploitation. As such it provides a rather extreme example of the consequences of lowland intervention in high mountain systems, consequences which might be avoided in the Himalaya. For this reason, as well as personal familiarity, the Andean material will be particularly relied on in emphasizing lowland-highland competition.

This competition is an important process in the tropical high mountains. Lowlanders are increasingly interacting with highlanders in an attempt to influence their actions and get greater access to their resources. The two groups appear to not only have rather different perceptions of one another, but also of the highland environment. As a consequence, considerable differences exist regarding an appropriate or adaptive solution to environmental problems and opportunities. When differences arise lowlanders have distinct competitive advantages in obtaining certain highland resources, and are exercising these more and more frequently.

The lowland-highland dichotomy is, of course, an over-simplification and in itself not sufficient to conduct an analysis. However, this paper focuses upon human groups engaged primarily in self-sufficient agricultural and/or pastoral activities. These groups are not only most prevalent in the high mountains but also depend directly upon the quality of the local environment in order to make a living. In turn, they are capable of exerting great influence on the environment since they use and modify it daily. Such groups frequently have been forced into living a rather marginal existence such that deleterious change can seriously alter the effectiveness of their adaptive system. In addition, these groups are the target of planned change aimed at making them more progressive and productive. Since they are socially and culturally different and often in ways little understood by lowland planners, the possibility of introducing deleterious change is all the more probable.

Highland subsistence groups at the individual household and community level appear to be the primary adaptive units interacting with the environment. They are also the units most sensitive to environmental change, and those in which variability in response can be observed most clearly. While planners frequently focus on the region as a unit of investigation, aggregating information at this level frequently obscures the precision with which adaptive responses may be understood. It follows that planning at the regional level may exacerbate the adaptive problems that the policy intended

to deal with. When lowlanders or lowland systems are referred
to, it will be done in a limited context. I am concerned
only with those individuals, institutions, or groups which
come from lowland areas into the high mountains and directly
influence aspects of the highland human/environmental system.

Turning to the potential contribution of this paper, an
additional summary on either high mountain human biological
or social systems is currently unnecessary. There is, however,
some value in attempting to integrate human biological and
cultural responses by noting similarities in their functional
capabilities of responding to high mountain environmental
conditions. Likewise it seems valuable to expand the analysis
and consider not only responses to the local biophysical
environment but also to the social environment within and
beyond the community. The primary goal of the paper, however,
is not simply to expand the scope of the analysis, but to
integrate existing generalizations into a conceptual framework
consistent with ecological and evolutionary theory.

Theory and models from theoretical ecology provide
hypotheses that assist in identifying critical parameters and
their relationships. When that is done the theory and models
allow some degree of prediction as to how variation in a
specific parameter or relationship will affect other aspects
of the adaptive pattern. These are fairly simple kinds of
prediction, which is not to say that they are also intuitively
obvious. The operational use of theoretical ecology in the
empirical study of the behavior of actual systems aids greatly
in identifying the critical variables contributing to the
overall adaptive pattern.

Hypotheses, deductively derived from theoretical ecology,
lead one to expect certain characteristics in the adaptive
pattern under given environmental conditions. Furthermore,
these hypotheses can be used to assess, in a qualitative
manner, the effectiveness of an existing adaptive pattern
should conditions change. When such generalizations are
applied to a knowledge of realistic behavior in high mountain
systems, the effects of a specific change can be qualitatively
evaluated. As a result, one is able to approximate the
adaptive limits of a response system and hence establish
broad guidelines for assessing the consequences of change.
In this manner, a conceptual framework is provided which may
be of use to administrators of planned change.

Models serve as a means of organizing information and
generating predictions of certain regularities. They can
operate on a variety of levels. As Levins (21) has pointed
out, models cannot simultaneously be general, realistic, and

precise. The present framework, in both a theoretical and analytical sense, will therefore attempt to emphasize generality and realism in providing broad guidelines for viewing change. Precise predictions of the effects of change on variables in a specific system requires an intimate knowledge of local conditions and historical processes.

In summary, this paper is oriented toward determining the adaptive pattern employed of self-sufficient groups residing in high mountain regions. High mountains are defined as regions above 3000 m, and emphasis will be upon the tropical and subtropical ranges, especially the Andes and Himalaya. Adaptive units under consideration are the individual, household, and community. Responses made by these units will be assessed with regard to their effectiveness in adjusting to both local environmental conditions and to lowland influence and competition. Limitations in the data and possible biases in the analysis have been pointed out. In spite of this, and the fact that deductive generalizations from theoretical ecology are imprecise and oversimplified, they serve as a starting point in understanding what constitutes inappropriate change in mountain regions.

The manner and form of exchange also has influenced dependence upon regional networks (19). Historically in the Andes and Himalaya, exchange and integration at the community level was achieved through kin and fictive kin relationships. With increased commercialization, need for cash has placed a growing emphasis on households to produce goods or services having a high cash value in the market.

Even relatively self-sufficient groups have rarely been completely isolated from regional events, large-scale religious movements, and outside political influence. Hindu penetration into the Himalayan foothills commenced sometime after the 12th century. By the 18th century the Hindu Gorkhali's control extended over a vast region from Kashmir in the west to Bhutan in the east (22, quoting 23). The more recent Chinese takeover of Tibet and the Indian government's influence over extensive southern Himalayan regions suggests the continuing impact of lowland ideology and control on mountain systems. These events, of course, have parallels in the Andes. With the colonial introduction of European religious and secular patterns, sweeping changes took place in highland human systems. Their degrading effects are most clearly noted in the substantial decline in Andean populations from preconquest levels (24).

Although past biocultural adaptations as well as regional and national integration all doubtlessly contribute in

significant ways to the characteristics of mountain communities
(3), the capacity of groups to adjust to these conditions
depends upon their adaptive potential within the limits of
the local environment. ' I will, therefore, focus attention
upon the environmental conditions of the local habitat, as a
first step in understanding the highland human adaptive
pattern.

Characteristics of the Environmental System

Several papers in this volume have described physical
and biotic characteristics of mountain environments. Most
have emphasized the extreme complexity of conditions which
exist at higher elevations. This makes generalizations
difficult. Mountains do not share broad environmental
commonalities in the way tropical forest or grassland biomes
do. Instead they are topographically diverse land masses
which create an increasingly unpredictable climate with
increased elevation and where most things occur on a sloping
relationship to one another.

Brush (19) has defined four climatic zones which consistently occur in high tropical mountain ranges: "(1) warm
valleys that are either dry or moist depending on their
relation to the prevailing rain-bearing winds, (2) temperate
intermediate valley areas, (3) cool high valley areas that
experience frequent nightly frosts in the tropics and seasonal
snow cover in mid-latitudes, and (4) arctic high valley areas
with permanent snow and glaciers" (p. 126). A succession of
vegetational zones or ecozones also follow an altitudinal
gradient, being influenced by climatic variables, edaphic
conditions, floristic history, and human modification (19,
quoting 25-28). The rich diversity of ecozones is especially
emphasized in tropical mountain ranges.

Interrelationships between the physical and biotic
environments in high tropical mountains have been extensively
covered in the literature. Summarizing this information the
following environmental stressors should influence the
adaptive responses of most tropical high mountain plants and
animals (29).

1. Reduced partial pressure of oxygen and carbon
    dioxide. Low absolute vapor pressure; high
    background radiation.

2. Rugged topography and poorly developed soils;
    marginal availability of certain nutrients.

3. Low temperatures with pronounced diurnal variation; frequent and intense frosts which can occur in any season.

4. Irregular monthly distribution of precipitation which can appear in various forms: rain, hail, snow. For areas having a lengthy dry season, droughts are not uncommon.

In addition to the above potential stressors, five physical and biotic environmental factors appear to be of critical importance in shaping human adaptive responses. These are relatively high degrees of: 1) environmental heterogeneity and 2) unpredictability, 3) low primary productivity spread over wide regions, and 4) high environmental fragility accompanied by a 5) downslope flow of materials. These factors are presented in some detail by Thomas and Winterhalder (29) and Winterhalder and Thomas (30) in reference to the Central Andes. What follows is a summary of this information as it is relevant to high tropical mountains in general.

Heterogeneity and Unpredictability

As mountains protrude high into the layer of atmosphere containing most weather phenomena (the troposphere) and divide continental air masses, climatic harshness, variability and unpredictability is increased. Variability in temperature, wind, humidity, precipitation, evaporation, insolation, soils, and geological substrate in turn produces a biotic environment which is heterogeneous in time, space, and pattern. With an increase in elevation, atmospheric pressure and oxygen tension decrease in a predictable manner.

Temperature shows pronounced diurnal fluctuations in a regular pattern, although extremes such as intense frosts, are generally unpredictable and are critical factors for vegetation development in the highlands. Fluctuation in precipitation follows a similar pattern on a larger time scale. While most high tropical mountains experience regular seasonal fluctuations in precipitation, the onset of the rainy season as well as the quantity and form of precipitation can vary considerably from year to year and locality to locality. Environmental heterogeneity and unpredictability are exaggerated in rugged, uneven topography, with varied exposure, wind patterns, up- and down-valley winds, and cloud cover; differential heating, which in turn effect convective storms and rain fall patterns, results.

Exaggeration is increased when climatic variability is considered with heterogeneity in soils, land forms, and geological substrate. A finely scaled mosaic of climatic conditions, topological features, and soils which result in considerable variability in vegetational forms appears to which the high mountain plants must adjust. They must, therefore, be capable of surviving both the frequent as well as the infrequent and rigorous perturbations. For the human groups who utilize and modify the highland vegetation, conditions conform to a patchy environment, where patches vary in size and pattern along altitudinal gradients, and from one valley system to the next.

Low Primary Productivity

A low energy capture rate by the plant community, characteristic of mountain environments, can influence the productive potential for human food and fuel. Few measurements of primary productivity have been carried out on plant communities in these regions, but estimates from the Central Andes suggest that energy production is low in relation to biomass (31) and dispersed over wide regions. Much of this productivity is accumulated beneath the soil surface and is consequently not available to grazing animals. Partial vegetative cover, especially on slopes, and dwarfing and slow growth of highland plants contribute to low productivity. Irregular environmental stressors such as intense frosts and droughts create unpredictable fluctuations in productivity from year to year.

Fragility and Downslope Flow

It seems paradoxical that high mountain systems which emphasize a high degree of resiliency should be described as environmentally fragile. Two factors, however, support the application of the term to human occupied areas. First, many highland plants are low and compact with extensive root systems taking advantage of the warmth and microrelief of the earth in order to provide protection against dessication, frost, and wind; but they consist of only a thin vegetative buffer to the frequently thin and immature soils. This becomes especially significant when such plant communities are located on slopes, since actions such as overgrazing, cultivation, and road construction can destroy portions of this vegetative layer and drastically increase the downslope flow of soil and substrate. Once this has occurred to a sufficient degree, the downslope vegetation, soil, and water quality of areas receiving the flow become effected, and system characteristics can become degraded for long periods of time over extensive areas. Precipitation falling onto steep topography with shallow eroded soils is subject to rapid runoff. During a heavy rainfall, Mann (31, quoting 32) estimates that 60% of Andean highland rain becomes runoff.

Adaptive Generalizations

Environmental conditions in high tropical mountains have been described as highly variable. A principal adaptive problem would, therefore, center around being able to determine when and where environmental problems and opportunities would occur; that is their predictability. As used here, predictability refers to the probability with which the adaptive unit can anticipate environmental opportunities or problems as they occur in time and space.

Justification for characterizing environments along a continuum from predictable to unpredictable is provided by Shugart *et al.* (33) as outlined below. Despite different approaches taken by Margalef (34), Child and Shugart (35), Holling (36), and Slobodkin and Rappaport (5) in studying the behavior of natural systems, their analyses converge on a general continuum of strategies:

1) "Systems strategies that represent an accomodation to change, emphasizing resilience and flexibility"; and

2) "Strategies that emphasize efficient responses to inputs, maintaining an equilibrium, and minimizing fluctuations." (33, p. 1718)

In stable conditions, stable adaptive responses will emphasize efficient use of a fairly constant group of resources. In unpredictable conditions emphasis is more upon flexible and effective utilization of a wider range of resources, resulting in resilient responses (Figure 1). These two distinct properties of adaptive systems have been examined in detail by Holling (36).

The utility of a predictability measure which provides a qualitative and comparative characterization of environmental conditions lies in the various levels of organization to which it can be applied. Deductive generalizations concerning appropriate responses can, therefore, be tested in a variety of adaptive systems; from ecosystems to physiological and genetic systems. This is important for determination of the adaptive pattern in high mountain regions since one must look for consistencies in a number of different adaptive systems and organizational levels.

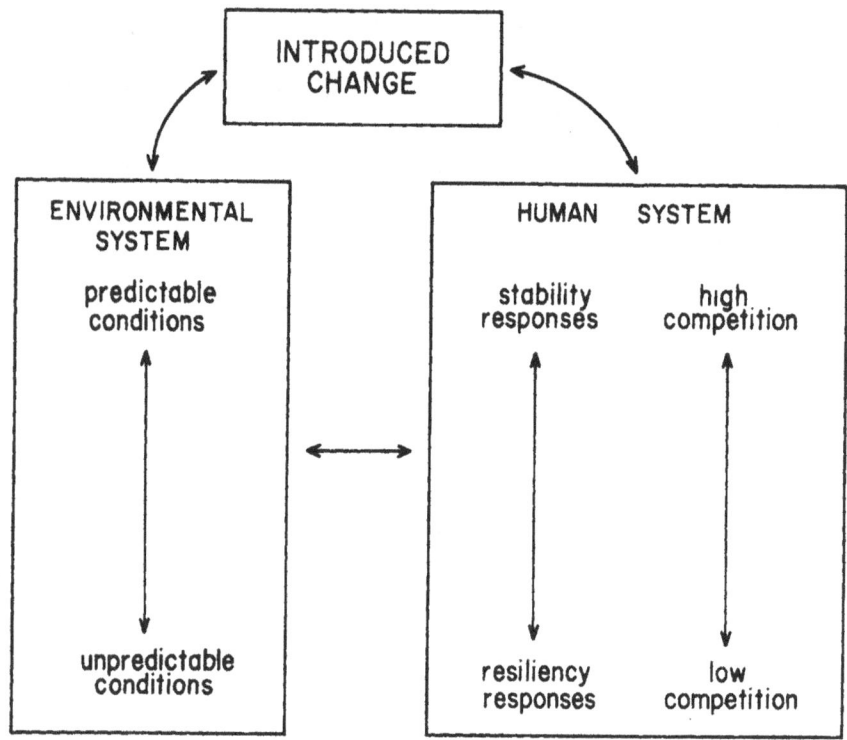

Figure 1.   Generalized diagram of dynamic interaction between
             introduced change and the human/environmental sys-
             tem. As environmental conditions become increas-
             ingly unpredictable, ecological theory suggests
             that human responses which emphasize greater
             resiliency and lowered competitive ability should
             predominate.

Since real systems of which man is a part are undoubtedly a mixture of both predictable and unpredictable conditions (37), it is possible to order conditions encountered by an adaptive unit along a continuum, and hypothesize general responses to them. This also permits a comparison of responses to diverse conditions with similar characteristics such as those originating from the physical, biotic, and social environment.

Adaptive Patterns of Flora and Fauna

Emphasizing the heterogeneous and unpredictable nature of many of these stressors, theoretical arguments would lead one to expect adaptive responses which emphasize a high degree of flexibility. This flexibility or resiliency could be genetic (ecotypic varieties or subspecies) or be achieved by phenotypic plasticity in physiological and behavioral systems of adjustment. Billings (38) has pointed out that both these mechanisms of adjustment are found in alpine flora; and ecotypic varieties are common among high altitude cultigens (39, 40, 41). To date, however, most data available on vegetation in the tropical high mountains are descriptive, and insufficient attention has been given to understanding the physiologic and genetic bases of adaptive mechanisms.

Although little information is available on faunal adaptations, several behaviors of Andean species, however, suggest a flexible adaptive capacity. Mammalian and avian species which are found along an altitudinal gradient emphasize less specialized strategies at higher elevations. Movement to high altitude may mean a prolonged period of development or a reduction in the number of offspring per year (31).

In a fluctuating and variable environment, an organism's potential for mobility is frequently important in avoiding environmental problems and taking advantage of patchiness. There are exceptions to this, however, especially among small mammals that are able to adjust within a single patch. Based upon considerations of species diversity, Pearson and Ralph (42) conclude that many small mammals respond to the Andean highland environment as though it were predictable and non-rigorous. Nevertheless their population density and biomass is relatively low. Bird biomass (g/hectare) is four times greater than that of small highland mammals suggesting that birds can utilize this environment more effectively than mammals of similar size; e.g., flight would generally decrease time and energy spent in searching for productive patches that are widely separated in space. As an animal becomes larger, the possibility of supporting itself in a single patch is progressively diminished. Koford's (43) studies of vicuna, for instance, indicate that a one-male band (several females and young) graze a territory ranging from 8 to 40 hectares.

Physiological adaptations to the low partial pressure of atmospheric oxygen, or hypoxic stress, have been summarized for native high-altitude animals by Bullard (44) and Folk (45). Low oxygen tension significantly interferes with physiological function of lowland mammals above 2500 m (46), necessitating compensatory responses to this persistent and regular stressor. Genetically isolated high altitude species show neither elevated red blood cell production (as measured by hematocrit ratio and hemoglobin concentration), nor a rightward shift in the oxygen dissociation curve, as do lowland species exposed to high altitude (see chapter by Grover in this volume for an explanation of these factors). Bullard (44) states that elevated hemoglobin and hematocrit ratio do not appear to yield any selective advantages and in fact may impose serious hemodynamic consequences. Several studies (47, 48) have shown that a rightward shift in the dissociation curve has favorable effects only in the low to moderate hypoxia range; there is a detrimental effect with severe hypoxia.

Distinctions between these two types of responses to hypoxic stress are of interest, since some species living at high altitude conform more to the lowland response than to that of genetically isolated high altitude animals. Exceptions are those species which maintain continuous genetic admixture from adjacent lowland areas (49). Humans perhaps fit in this category--although the evidence is hardly conclusive for native high altitude residents (50). High altitude human populations have probably never been genetically isolated to the extent that other native species have, nor are they restricted to a given altitude (51, 52, 53). Utilization of multiple resources at different elevations and interzonal exchange requires a physiological system which is flexible enough to respond to a range of hypoxic conditions. Human exposure to low partial pressure of oxygen follows a pattern much more variable than has been previously depicted in the literature; hypoxia is neither fixed in intensity nor regular in occurrence for organisms which utilize altitudinal gradients. This suggests that genetic adaptations which allow organisms to conform to a specific set of environmental conditions, such as an ecotypic variety to severe hypoxia, become increasingly inappropriate as the organism must move over broad areas, altitudes, or zones to meet requirements. For organisms exposed to both a variety of environmental conditions and rapid change, emphasis on genetic adaptations permitting a wide range of phenotypic plasticity would seem more appropriate

It is, therefore, expected that the human adaptive pattern would demonstrate considerable plasticity in high tropical mountain regions.

The Human Adaptive Pattern

While commonalities in adaptive pattern between a number of high mountain life forms have been suggested, our understanding of adaptations to the heterogeneous and rapidly changing environments of these regions is yet scant. Nevertheless, it is possible to examine this pattern in somewhat more detail by turning to studies on human adaptation to mountains. Not only have both biological and behavioral adjustments of people received more attention than other highland species, but the conditions to which such groups are exposed are possibly more diverse and transitory, making mountain groups particularly appropriate examples.

Humans living in the tropical high mountains are frequently exposed to diverse conditions. High levels of insolation and heat by day and cold by night may be experienced. In regions where seasonal variation in precipitation occurs, relative humidity and level of insolation change significantly in the course of the year, and factors affecting the transmission of infectious disease also shift under such conditions. Because the seasonal pattern of freezing temperatures and rainfall exert an influence on high-altitude agriculture, work levels change from periods of low activity to those requiring sustained hard labor. This occurs at different altitudes and under different hypoxic conditions. Nutritional intake, quantitatively and qualitatively, shifts in the course of the annual cycle. And droughts, hail storms, early snows, intense frosts, and crop and herd pathologies can cause food production to fluctuate in an unpredictable manner from year to year. Influences and competition from other human populations can lead to more long-term changes in environmental conditions. Such changes alter existing relationships with the environment, necessitating a reorganization of adaptive responses.

Responses at both biological and cultural levels are needed in this multiple stress environment. While on some occasions organisms need only adjust to one stressor, more frequently a number of environmental problems must be counteracted at once or within a close period of one another. Many adaptive responses, therefore, have multiple functions and thereby are capable of counteracting more than one problem; this is referred to as "cross-adaptation" (54).

Adequate clothing, for instance, is a positive cross-adaptation to both potential cold stress and low energy availability; in keeping an individual warm it also reduces the cost of thermogenesis. A somewhat inferior response to cold stress could be the maintenance of a high level of subcutaneous fat (assuming this is possible in an energy deficient environment). Fat is cross-adaptive in a negative way because additional body weight in the form of fat increases the oxygen and energetic cost of activity. In turn, an individual's ability to engage in strenuous work would be reduced and both hypoxic and hypocaloric stress accentuated. Fat deposition is a less reversible response than addition of clothing.

Unfortunately, very few empirical data on cross-adaptation to transient multiple stress conditions exist which can be applied to high mountain groups (55). Nevertheless, the concept of a highly interconnected web of positive cross-adaptive responses is important. Although specific responses for each condition are plausible, it should be kept in mind that a response in one direction can compromise the organism's ability to react to other conditions. Therefore, in a multiple stress environment, we might expect biological and cultural selection for positive cross-adaptive responses. When this is not possible and responses are cross-adaptive in a negative way, they should be rapidly reversible once the stressful period has passed. This argument is speculative but it is put forth since it may influence the characteristics of a response to a given condition and whether the response is evaluated as relatively adaptive or maladaptive. While small body size and retarded growth of many high altitude groups can be classified as maladaptive (at least compared to the Western European and North American standards), it is possible to show that, up to a point, reduced stature and growth rates lower a population's energy requirements and do not interfere with response to hypoxia, habitual work, and, if clothing is worn, cold (56).

Since biological and cultural adaptive responses contribute to this web of adaptation, it is inappropriate to consider either one or the another as operating in isolation. The presence of some rather unique biological characteristics among high Himalayan and Andean human populations suggests that cultural responses are sometimes unavailable (as in the case of hypoxia), not completely effective in buffering the environment (cold), and occasionally fail (crop loss resulting in undernutrition). While biological and cultural adaptive responses obviously interact, they do have very different capabilities in solving environmental problems. In some cases it is more effective to modify or buffer the problem before it effects the organism and disrupts its physiological homeostatic balance. In others quite the reverse is true.

There are conditions that are either very costly or impossible to block, hence morphological and physiological solutions appear more appropriate. In any case these two broad types of response both contribute to the adaptive success and therefore should be expected to share functional similarities in their adaptive pattern.

Specific human biological responses to hypoxia, cold, undernutrition, and other high altitude stressors have been recently summarized (14-17). In terms of an adaptive pattern, it is of interest to review the degree to which the biochemical, morphological, and physiological characteristics of high altitude populations are attributable to phenotypic plasticity, and hence more flexible to diverse and changing conditions. Although many of these characteristics have been described as genetic adaptations (57-60), recent evidence suggests that conditions encountered in the course of the growth and development period may be significant in shaping adaptive responses to cold (61) and hypoxia (62).

Frisancho and others (63), for example, have compared maximal oxygen consumption values (aerobic capacity) of three young adult samples: 1) high altitude Peruvian natives; 2) lowland migrants who moved to high altitude with their parents at ages ranging from 2 to 16 years of age; and 3) newcomers to high altitude. While the second sample appeared to have a higher admixture of European genes than highland natives, oxygen consumption values of these two groups were nearly identical; newcomers expectedly showed rather low values. Moreover, individuals who moved to high altitude as adolescents had significantly lower maximal consumption values than those who came as young children. Maximal oxygen consumption is a measure of the oxygen transport system's capacity to provide the metabolizing tissue and is an effect of many and specific biological responses operating together. As such it is a measure of an individual's ability to perform strenuous, prolonged work. In summary, genetic factors obviously do influence adaptations of high altitude groups, but studies which demonstrate that these adaptations have a high heritability component have not appeared. Lacking specific genetic association with responses to high altitude stressors, support that high altitude human populations are genotypic ecotypes awaits confirmation.

Behavioral and cultural aspects of the adaptive pattern are more easily identified and easier to verify in humans than biological ones. Humans, in adapting to rapidly changing conditions throughout evolution, have relied primarily upon behavioral flexibility, including the ability to predict and modify environmental conditions by high emphasis upon technology,

Table 1. Human production zones in three mountain regions. Adapted from Brush 1976, p. 128 (19)

Production zones	Southern Switzerland		Peruvian Andes		Central Nepal Himalaya	
	Approximate altitude range (m)	Products	Approximate altitude range (m)	Products	Approximate altitude range (m)	Products
Low altitude	<1000	Fruits, especially vineyards	<1500	Sugar cane, coca, fruit, rice	<1500	Rice, fruit
Mid-altitude	1000–2000	Cereals, hay, gardens	1500–3000	Cereals	2000–3000	Cereals, tubers
Mid/high altitude	2000–2300	Forest	3000–4000	Tubers	3000–4000	Forest
High altitude	2300–3000	Pasture	4000–5000	Pasture	4000–5000	Pasture

Table 2. Predominant peasant crops and livestock in three mountain regions. Adapted from Brush 1976, p. 129 (19)

	Barley	Wheat	Maize	Rice	Potatoes	Other tubers	Sheep	Cattle	Goats	Cameloids
Alps	+	+			+		+	+	+	
Andes	+	+	+		+	+	+	+	+	
Himalaya	+	+	+	+	+	+	+	+	+	+

information transfer, and cooperative responses. Although all human groups use such strategies, they do so in different ways and to different degrees. Some behavioral and cultural solutions to similar conditions will be expected to differ between highland and lowland groups.

Figure 2 indicates three interrelated types of responses which appear to constitute the core of the highland human adaptive pattern. These are reliance upon a multiple resource base, considerable emphasis on the prediction of local conditions, and a high degree of behavioral flexibility at the individual, household, and community level. These are in turn related to five more specific response types, all of which contribute to the maintenance of continuous access to essential resources, and presumably human biological well-being.

Multiple Resources

In adapting to environmental heterogenity and unpredictability, dependence upon a multiple resource base (agro-pastoral) is a production strategy which both takes advantage of environmental opportunities and minimizes risk from perturbations affecting single resources (18, 64). When an extensive altitudinal gradient is available, a diverse range of animals and crops are produced in different zones. While most of investigators have noted the highland emphasis upon multiple resources, Murra's (65) ethnohistorical work on "vertical control" in the Andes most clearly illustrates this point. Vertical control refers to the observation that, traditionally, human systems have maintained access to a large number of altitudinal ecological zones through kin relations and other social relationships in order to achieve a reliable and integrated resource base.

In the Alps, Andes, and Himalaya, Brush (19) has pointed out that a succession of four human ecozones are distinguishable based upon both products and production regimes (Table 1). The mixed agro-pastoral base in all three mountain ranges share a number of similar crops and animals (Table 2).

Predictability

Adaptive responses by organisms living in high tropical mountain environments are generally effective in reducing potential unpredictability for some of the conditions encountered. Humans are rather successful in this respect and employ two basic pathways. One is to anticipate environmental problems and opportunities through knowledge of their characteristics. Much attention must be given to acquiring and exchanging

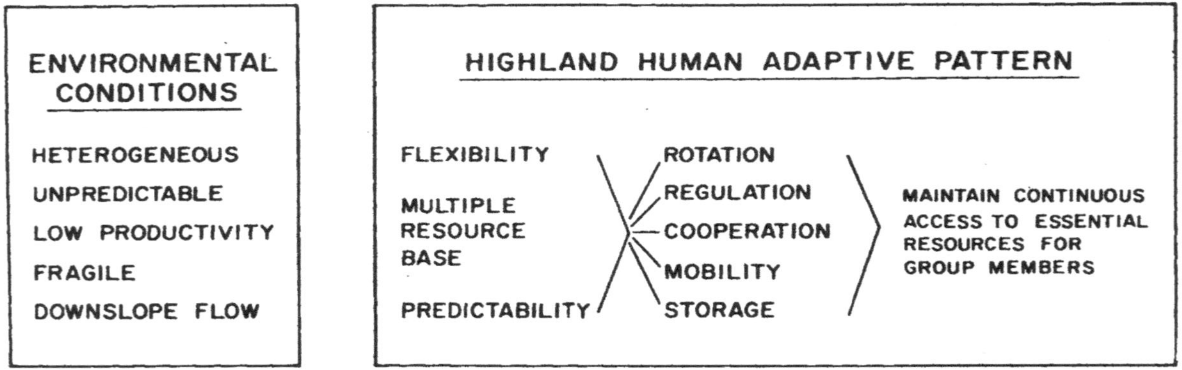

ENVIRONMENTAL CONDITIONS

HETEROGENEOUS
UNPREDICTABLE
LOW PRODUCTIVITY
FRAGILE
DOWNSLOPE FLOW

HIGHLAND HUMAN ADAPTIVE PATTERN

FLEXIBILITY
MULTIPLE RESOURCE BASE
PREDICTABILITY

ROTATION
REGULATION
COOPERATION
MOBILITY
STORAGE

MAINTAIN CONTINUOUS ACCESS TO ESSENTIAL RESOURCES FOR GROUP MEMBERS

Figure 2. Predominant characteristics of high tropical mountain environments and human adaptive responses to them.

information about the local environment because of its complexity. A second pathway is modification of conditions in ways that make the environment continuously productive and hence more predictable.

Such modification, however, has limits. The inability of most groups to substantially alter climate and terrain is a significant constraint in highland areas. Also, if modification is to be of long term adaptive significance, it should not lead to a substantial increase in environmental deterioration and ultimately greater unpredictability. For example, root crops are planted on slopes for better drainage; but because soils must be dug deeply in order to plant and harvest such crops, significant downslope soil loss is expected, especially on the steeper slopes and when fallow periods are shortened.

In a very heterogeneous environment where patches and ecozones have differential utility to human groups, travel and transport across unproductive ones are costly. Natural processes on altitudinal gradients, according to Beals (66), tend toward diversification of ecozones because of specialization and competitive exclusion. In contrast, as a user of the whole gradient, this large human mammal tends to modify, redefine and reduce the zonation. The tendency is in the direction to alter natural processes, perhaps reflecting the human adaptive penchant for control. In addition, since humans in order to have a resilient system must depend on multiple aspects of the altitudinal gradient, it reduces between patch travel time to create somewhat larger patches (all of which are used or visited at different points) (67).

Although a rather consistent group of domesticated resources are produced within each of the four human ecozones mentioned above (19), a considerable number of variants of these resources are suitable for the diverse patches within each zone. Ugent (40), for instance, reports over 400 different varieties of potatoes in the Andes, suggesting the considerable attention given by indigenous plant breeders to developing crops suited for different conditions. Thus, if a highland native knows how to assess the conditions in a prospective field, has the proper seed, and can recruit enough assistance, a patch normally suited for grazing or forest resources can be made more productive by cultivation. Furthermore, as more "natural" patches are made suitable for cultivating varieties of several crops, a zone becomes less heterogeneous from the viewpoint of its productivity to the human community. The extent to which such modification increases environmental predictability is reflected by the degree to which increased productivity acts to 1) buffer the

variable effects of environmental problems that existed in
the premodification stage, and 2) does not add a new set of
unpredictable conditions. Modification, therefore, frequently
entails utilizing a variety of cultivars which compliment one
another in tolerance, requirements, scheduling, and production
(29). A similar complementary pattern exists between cultigens
and domestic animals.

Herding, typical of the higher zones, is a modification
which simultaneously decreases the environmental heterogeneity
and unpredictability. Domesticated herbivores can more or
less efficiently utilize many of the patch types which cannot
be used for cultivation and thus have little direct utility
to humans. Also, suitable grazing areas can be expanded by
deforestation, burning or, in drier areas, diverting streams
perpendicular to a slope and letting the seepage water the
area below. Unlike cultigens, which are dependent upon
conditions in a restricted area, herbivores can utilize
pasture sources over a wide area. They can be moved and are,
therefore, less influenced by local climatic disturbances.
Because of this they constitute a more reliable human food
source. While an environmental perturbation may destroy a
large portion of crops in a local area, domestic animals can
survive by relying on this mobility and on greater endogenous
food reserves (29), both of which seem to be of critical
importance in introducing a degree of stability to fluctuating
and unpredictable conditions.

Environmental modification by cultivation involves
intensive utilization of relatively small and often dispersed
areas up to a given, limiting, altitude. Pastoralism, which
frequently occurs above agricultural limits, is, however, an
extensive utilization system which can be practiced in less
productive patches where herds forage widely concentrating
energy and materials. Animal protein, hides, and fiber which
are limited in the lower tropical zones can be produced in
abundance higher up, insuring participation of even specialized
pastoralists in interzonal exchange.

The positive cross-adaptive aspects of relying on herd
animals is apparent. Not only do they provide a food source
from generally unproductive zones, but they also contribute
to solutions to nutrient depletion of the soils (dung) and
the buffering of cold stress (fuel and clothing). Furthermore,
they facilitate the transport of bulky items. All of these
factors involved in the agro-pastoral base contribute to
gaining access to material and energy flows in and beyond the
local environment. As a result a number of environmental
constraints become both less disruptive and more predictable,
and environmental opportunities are expanded.

Flexibility

Although overall environmental predictability and
productivity appear to be increased as a result of well-
integrated subsistence strategies, unpredictability remains
an important factor in the production of specific resources,
especially cultigens. Individual crops can be lost throughout
the growing season, and human time and energy put into them
remains uncompensated. Given such conditions, it is expected
that high flexibility in abandoning one set of responses and
shifting to a new environmental opportunity is adaptive.
When opportunities are widely dispersed in space, rapid
mobility is important. Cooperation between households is
also of high adaptive value. Frequently a household must be
allowed to link up with another ongoing productive activity
and share some of its benefits.

Specific Response Types

All of the specific response types which contribute to
the broader adaptive pattern· function to coordinate problem
adversion and access to opportunities as they come and go.

Rotation. In a region where resources in any one area
are limited, and sustained efforts by organisms to fulfill
their needs can seriously degrade that area, rotation becomes
an important adaptive policy. Rotation of crops and herds to
prevent depletion of nutrients, overgrazing, and erosion is
also typical in high tropical mountain regions.

The overseeing of the agro-pastoral rotation and other
activities becomes the responsibility of village guardians
who are also selected on a rotating basis so that this
responsibility eventually falls on all household heads in the
community encouraging group cooperation (18). "The rotational
system of office holding symbolizes the tendency for total
community involvement which characterizes all aspects of
alpine village life" (18, p. 541). Such rotation discourages
permanent or long-term establishment of guardians who might
use their position to acquire a disproportionate share of
local resources.

Regulation. In an adaptive sense, tradition functions
to pass on consistent patterns of behavior which individuals
would find dangerous or complicated to discover on their own
(68). Viewing this in terms of regulation in highland
communities, people of course can behave in ways capable of
creating disruptive conditions, both socially and environmentally, that cannot be permitted if the community is to
persist. Controls are, therefore, established to elicit and

Table 3. Characteristics of land used communally and privately. Adapted from Netting 1976, p. 144 (71)

Nature of land use	Land tenure type	
	Communal	Private
Value of production per unit area	Low	High
Frequency and dependability of use or yield	Low	High
Possibility of improvement or intensification	Low	High
Area required for effective use	Large	Small
Labor- and capital-investing groups	Large (voluntary association or community)	Small (individual or family)

enforce consistent behavior from members for critical activities.
While all communities have such regulations, in highland
groups these focus upon access to community-held resources, a
concern for their sustained yield, and the maintenance of
cooperation between households.

Rhoades and Thompson (18, summarizing 69, 70) provide an
example of such regulation among the Sherpa:

> Each year in early May the village officials concerned
> with the regulation of herding and cultivation are
> selected from among the villagers. Shortly thereafter
> a village assembly...is held, and it is decided when
> and how far from the village herders must move their
> animals. This legislation is written down and passed
> on to the (official), who must then administer the
> wishes of the people. The guardians must also oversee
> the cultivation and growing of crops near the main
> village and ensure that no one enters any field, even,
> his own, until the potato harvest begins.
>
> Sherpas have a separate set of officials...who protect
> the village forest lands. The (officials) have the
> right to fine anyone who violates any of the strict
> rules dealing with forest land. Any offenders are
> required to appear before the entire village population
> to be publicly fined and shamed. The fine may consist
> of beer or cash, depending on the severity of the
> crime. If beer, it is consumed immediately after the
> trial in a celebration in which the offender actively
> participates, another indication that hostility
> cannot be permitted to persist for long in an alpine
> setting.

Communal regulations function not only to promote
effective production of community owned resources, but also
to control access and utilization of them in order that they
not be degraded. In a fragile environment where unchecked
exploitation can cause rapid and irreparable damage resulting
in denial of essential resources to a portion of the community,
conservation ethic supplemented by enforcement appears
necessary.

Not all resources in highland areas fall under community
control. Netting (71) in a study of a Swiss alpine village
notes associations between land use and type of land tenure
type (Table 3). This indicates that private ownership of
resources prevails when land is more productive and predictable,
or can be made so through modification by a single land
owning unit. On the other hand, communal resources tend to

be those that are more dispersed and unpredictable and would both lose productive value or require increased labor input (as a result of duplication of effort) if broken into private parcels.

Increased population pressure on a restricted resource base can ultimately result in damage to the highland environment (18). Limiting access of offspring to household-controlled resources appears a common regulatory mechanism. Inheritance patterns for instance are one means of maintaining viable economic units (64). In the Alps, strategies of land use and sociopolitical organization require a constant outmigration which can reach 50% per generation thus controlling local population growth (3). Mercenary soldiery in the lowlands (22) produces a similar result for Nepalese highland groups. Substantial out-migration of young adults has also been reported for the Andes (72).

Regulation also applies to attempts to limit outside influence over community resources. Two principal explanations are given for this response (3). The first proposed by Wolf (73, 74) suggests that "closed corporate" communities result from outside political and economic demands on labor and resources. When demands are made from the outside, it is the community organization that intervenes. The community attempts to manipulate such interactions in order to minimize outside influence over local resources. When demands must be met the organization distributes the burden among households so that outsiders do not directly interact with these units.

A second explanation, based on Swiss alpine ethnohistorical evidence, focuses more upon local environmental constraints and subsistence requirements (71). Local autonomy and the corporate features of the community are viewed as an adaptive response to population pressure on a limited resource base-- land use influences land tenure patterns. In both explanations environmental predictability and response flexibility is increased for the "closed" community. Whether this is a response to the social or physical/biotic environment is of less concern than the positive cross-adaptative nature of the response.

Cooperation. Cooperative strategies are emphasized when 1) an individual unit is not able to respond sufficiently to environmental conditions, 2) a unit can be assured that another will assist when necessary conditions arise, and 3) the interaction is ultimately of benefit to both parties.

In high tropical mountain regions groups frequently neither have access to the entire range of regional resources nor do households within a community control all the resources they need (75). The unpredictable nature of environmental problems likewise presents conditions where a household cannot solely rely upon its own efforts or resources. Both of these conditions necessitate a dependency upon other group members as well as individuals beyond the community.

While such dependency can take a variety of forms, reciprocal relationships based on kin and fictive kin are frequently emphasized in more traditional mountain communities (22, 76, 77). One advantage of forming cooperative unions with relatively strong and long-lasting bonds, such as kin cooperation, is the extent to which generalized reciprocity can operate. Repayment in kind or within a particular time period is not as stressed as it would be under a more formal cooperative arrangement.

Such an arrangement provides considerable flexibility for a household which encounters a sequence of bad luck, since it demands only that the household contribute to the cooperative unit what it can afford and when. Conversely, to deny the household continued assistance suggests that it would ultimately have to leave the community. As indicated, cooperative exchanges in highland communities are based upon an equalitarian rather than a hierarchial social organization. This does not deny differences in wealth and power between households, but suggests that these are quantitative rather than qualitative (3).

In the Andes, subsistence activities such as field preparation, harvest, shearing, and slaughtering frequently require more assistance than a household has available. Since timing and coordination of activities are important aspects of the subsistence pattern, the household must gain access to labor when the need arises especially at planting and harvest time. Likewise, when an opportunity appears unpredictably, it is necessary to rapidly assemble a group and take advantage of it.

As Gero (78) has noted, scarcity of labor is influenced by low population densities, a dispersed settlement and exploitation pattern, subsistent activties which require high energy inputs by humans, and restricted in-migration coupled with an out-flow of young adults. One advantage of securing assistance from within or between kin groups is that such groups are held responsible for honoring the reciprocal relationship. This is not always the case when contracts are made at the individual level. Unlike other forms of repayment reciprocal labor exchange reaffirms social alliances and thereby reinforces household interdependencies. In addition,

it enables participants to collect and exchange first hand
social and environmental information on crops and animals,
modes of production, and microzones from a variety of households.
With such high dependency and continuous exposure to members
of other households it, therefore, becomes difficult to cheat
or manipulate the labor exchange system.

Considerable emphasis in the highland adaptive pattern
is placed on social solutions to environmental conditions.
Technological solutions, especially the material culture,
appear somewhat less impressive. It is possible to make
certain functional generalizations about conditions under
which "tools" would be heavily relied upon. For sedentary
groups who exploit resources in close proximity to their
homes or a storage location such as a barn, a wide array of
specialized tools may be accumulated and brought out as the
need arises.

On the other hand, when a group is more mobile (transhumant),
exploits widely dispersed resources, and cannot always anticipate
the conditions when specific tools are needed, certain limitations of this strategy become apparent. For instance, the
costs of transporting tools might outweigh their benefits.
Under such circumstances, it is expected that material technological solutions would be most appropriate for 1) those
environmental conditions that are relatively predictable in
time and space, 2) can operate without humans being present,
or 3) are useful in responding to a variety of conditions.
The last suggests that such technology would be taken with
individuals or households as they move.

In terms of high tropical mountain systems, consistent
technological responses center around the employment of
multiple cultivars suited for diverse microzones, the checking
of erosion by terraces and other structures, and the channeling
of the downslope flow of water for irrigation, etc. All of
these meet persistent environmental problems (low productivity,
downslope flow) or opportunities (power generated by flow of
water). Locations where they are appropriate are easily
identified and once established they can function effectively
for variable periods without human supervision.

Mobility. The continuous need to exploit different
zones and niches along an altitudinal gradient, to exchange
resources within the region, and to respond rapidly in spatial
movements as unpredicted problems and opportunities appear
underlines the importance of mobility (18). This in turn
demands that means exist to facilitate the movement of people,
their portable possessions, and the materials they exchange.
Mobility, therefore, represents a type of technological

response which permits adjustment to a variety of conditions.

The detailed regulations surrounding the maintenance and use of avenues of transportation in many more traditional mountain societies points up their importance. Herd animals that produce meat, hides, fiber, fertilizer, and fuel serve as transporters of cargo in the Andes and Himalaya along with related transportation technology such as ropes, saddles, containers or sacks, and clothing suitable for diverse conditions. Household possessions must be designed so that they are portable. Since transport animals have limits as to the bulk and weight they can carry, items which exceed their capabilities are excluded as are extremely fragile items which cannot be safely packed. I am not aware of any study which examines highland household possessions both in terms of their portability and how this effects their design and function but one would expect principles of cross adaptation to apply here.

Storage. Another broad technological response which is effective in a variety of conditions is the storage of resources. This is a particularly appropriate response in environments with seasonal fluctuations in productivity, varying levels of production from year to year, and where the preservation of foods and other items is possible. All these conditions are generally met in high tropical mountain regions. Freezing nighttime temperatures and warm dry days permit the dehydration of both vegetables and meat (26). This food preparation process not only significantly increases the period for which such items can be stored, but also results in a substantial loss of weight facilitating their transport. For instance, Andean dehydrated potatoes can be reduced by approximately 75% of their initial weight and kept almost indefinitely (79). Morris (80) has reviewed the extensive storage facilities available in Incaic times. Cheese making likewise eliminates the necessity of transporting milk. Hay (in the Himalaya) and cereals can be stored without much preparation. Animal fibers are woven into cloth and stored; in event of hardship this can be sold or exchanged (81). An accumulation of herd animals serves a similar function.

In a group where each household attempts to buffer unpredictable conditions through storage, and reciprocal relationships link households in different ecozones, stored resources can be used to redistribute gains and losses, supplementing reciprocal labor exchange. Thus the cumulative level of stored goods within a community is an indicator of resources potentially available for redistribution and the community's ability to buffer environmental perturbations.

Effects of Lowland
Competition and Influence

The human adaptive pattern discussed above constitutes
an integrated complex of responses which appear to be effective
in keeping critical variables in highland human systems
within their tolerance limits. Three general changes, however,
can degrade the effectiveness of the adaptive pattern. These
entail 1) increasing the unpredictability, frequency, duration
or intensity of existing environmental problems, 2) adding
new environmental problems, and/or 3) reducing the effectiveness
of the aforementioned adaptive strategies.

All of these factors appear to be contributing to the
collapse of highland human systems, as evidenced by extensive
downward migration and village abandonment. One lowland
perspective is to view such a process with relatively little
concern. The rationale underlying this is that highland
peasant groups are unprogressive, offer little to the development of national economy, resist change from governmental
agencies, and contribute to the deterioration of the highland
environment. If one is convinced of this, plans which
encourage migration to the lowlands in order to make highland
peasants more "productive" and to facilitate greater national
control of highland resources seem both sound and proper.
Thus reforestation programs, hydroelectric facilities, and
mining operations could all be pushed ahead without the
frequent resistance from local groups. While most planners
would agree that further deterioration of the highland
environment is to be prevented, no such concensus surrounds
what is best for highland human systems. Furthermore, with
so little information about these systems, it is hard to
evaluate with any precision the extent to which they are
actually being disrupted. Economic indicators are of little
value since production and exchange in these systems have
never been recorded. Public health indicators, such as
undernutrition, usually suggest extensive deterioration of
the system has already occurred. Village abandonment as an
indicator appears too late.

While there are indeed some advantages for lowlanders in
the depopulation of high tropical mountain areas, the costs
may be rather immense. Highland groups are yet fairly self-
sufficient and sustain themselves largely without services
from the national government. Furthermore, they produce
resources such as animal protein, hides, and fiber as well as
high-protein crops which are scarce at lower altitudes.
While lowlanders realize that some highland products are
important to the national economy, recommendations are
usually that these be produced on a larger, more commercial

scale; this usually entails capital and technological specialization. Before initiating such an action, however, it is
necessary to anticipate what effect restructuring production
patterns would have in forcing individuals out of highland
systems. Since out-migrants go to larger towns and urban
areas where they compete for opportunities with local residents
and demand government services, such information is critical
for lowlanders. For lowland systems that are ill prepared to
receive this downslope flow of humans, problems of greater
magnitude than those caused by the downslope flow of soils
are eminent.

Since checking population growth and increasing both
human productivity and land reclamation are major concerns of
nations containing high tropical mountains, it seems contra-
dictory that more concern is not given to preserving self-
dependent highland groups. It is doubtful that migrants from
low altitude to the mountains, not having undergone developmental
acclimatizations, could adjust to and work comfortably at
high altitude; a substantial reduction in lowlanders working
capacity is evident at high altitude. Biological and psychological discomfort levels from cold, diet, and isolation
would discourage many lowlanders from considering permanent
residence in these regions.

Assuming, however, that it were possible to recolonize
abandoned areas, or that highland groups were to remain where
they are in some altered form, cooperative networks and other
social responses are difficult to recreate. Agrarian reform
movements in the Andes, for instance, have assumed that the
formation of cooperatives would lead to the reestablishment
of an equitable sharing of resources among households. This
has generally not been the case (82). While technological
solutions can sometimes replace social responses to highland
conditions, these usually are at a high cost to the government.
Finally, and possibly of greatest importance, is the loss of
information concerning how to use the local environment when
a highland human system is significantly altered. This is
information which has been built up and passed on for generations. It will not be easily relearned by lowland migrants,
highlanders from other areas, or even agronomists and soil
scientists trained in the best institutions of the world.

Lowland Competitive
and Influential Potential

With increasing efforts by lowlanders to influence and
compete with highland groups, we should consider those
resources for which lowlanders would be expected to have a
competitive advantage. Concentrated resources such as salt

and mineral deposits are easily discovered by lowlanders;
highlanders have no particular advantage in being able to
anticipate when and where these will occur. Major trade
routes and markets act as concentrators of resources and
human productive activity. Concentrated resources also are
considerably easier to control than the dispersed and small
scale production of individual villages or households.

Lowlanders also have greater competitive abilities for
those resources which are more effectively obtained with
capital-intensive technological solutions, many of which
center around concentrated resources such as mining, production
of hydroelectric power, and irrigation over uneven terrain.
Creation of a modern transportation system serves as another
example of competitive superiority. Although vehicles are
not as flexible as animal transport in crossing mountain
ranges and linking diverse areas of importance to highland
systems, they do join locations which are of significance to
lowlanders. Roads are effective in channeling and intensifying
the flow of goods (including essential resources) and collapsing
animal transport exchange routes. Communities reorient
toward the roads, becoming more reliant upon the truckers and
the products they bring. In turn, peripheral communities
become cut off from the exchange network which is now controlled by individuals outside the community with little knowledge
or concern for their problems.

The outcome of competition for factors of production,
such as land and labor, is of critical importance to the
persistence of highland systems. In broad, fertile valleys,
a variety of crops and farming techniques can be employed
making them prime contested areas. Information as to how to
effectively utilize such an area gives the highlander no
particular advantage. In fact the utilization of mechanized
farm equipment requiring less dependence on labor, as well as
a better knowledge of the national economy, provides the
lowlander with a decided edge in effectively utilizing land
in such areas. Consequently, the highland human system loses
control of a highly productive region, and must rely on more
and more marginal areas with their greater environmental
unpredictability.

Competition for human resources shows a similar trend.
Wage labor offered by nearby mines or lowland labor recruiters
is enticing for those experiencing a relative scarcity in
material possessions. Whether one goes to work outside the
community or stays in the area, a strain is placed on the
effectiveness of reciprocal labor exchanges. Kinsmen not
engaged in wage labor must take on a greater burden, and a
point is reached at which some households cannot get the

cooperation they depend on. In this case, critical resources flow into the family below their tolerance limit and other alternatives must be sought; wage labor is one solution, leaving the community another. In either case the reciprocal labor exchange becomes further strained.

While the purchase of legal access to highland resources sometimes results from the lowlanders ability to use these resources more efficiently (broad fertile valleys), it need not always mean greater success of lowland responses. Given sufficient discrepancy between reserves of highlanders and lowlanders to buffer environmental problems, the latter can wait for an opportune time (a drought) when a highland system has been severely challenged. When such an occasion arises communities and households, despite their reluctance to allow outside control of resources, are sometimes forced to make concessions. The result is a further reduction in adaptive flexibility to meet future problems.

Since lowlanders are generally more familiar with the utilization of governmental laws and services, or in a position to influence them, these mechanisms can be used to enhance their competitive abilities. This aspect of the social environment is less predictable to self-dependent highlanders. Legal manipulation of land and resource titles, and a reliance upon enforcement agencies to back up claims, has not been uncommon in the Andes. Likewise political reforms and technological assistance programs can be used by the government to gain control over local resources and break down community autonomy. When such programs do not provide uniform services to all households, but favor certain families or segments of the community, they frequently act to the detriment of the group. For example, administrators of reform and assistance programs are accountable for showing progress (usually by some measure of human productivity) to their governmental agency. Although these individuals may be well-trained specialists in their area, they frequently are unfamiliar with both the local environment and people. They, therefore, must rely heavily upon families who 1) accept their presence in the community, and 2) can be instrumental in meeting agency goals.

To provide an extreme but not uncommon example, technological assistance is given to families with relatively large land holdings, since they can use it more effectively. This action generally results in production increases over previous levels, and, on paper, a successful assistance program. What is unreported is how this effects production of households (per capita production) not receiving assistance, and their ability to cope with future environmental conditions. It is possible, for instance, that conditions are altered to the extent that many households must sell their private land holdings and leave the system in spite of the optimistic reports sent to the government. In many cases the local administrator has no way of assessing damage done to the system since he has no background in its past operation. Nevertheless, by assisting already powerful families to increase their wealth and to operate somewhat independently from reciprocal exchanges, checks which the community traditionally might have used to maintain an equalitarian society become ineffective.

Another example of government influence on highland groups is education. Formal education which provides reading, writing, and arithmetic appears essential if one is to deal with the national economy. In spite of this however it must be seen, in part, as a competing information system oriented towards integrating citizens and communities into the national goals and values. In the Andes, educators are frequently not members of the highland community in which they teach. Just as the administrator of planned change, they are charged with influencing community members in a manner thought appropriate by national educators. Thus the learning of essential local information such as pasture types and soil indicators is replaced by national history, geography, and sometimes a foreign language. Furthermore, children are sometimes subjected to a value system which characterizes local customs as inferior and orients them to opportunities beyond the community. The essential point is that formal education when controlled by outsiders, and especially lowlanders, provides a world view which is frequently incompatible with that of the community. Furthermore, it adds little information which is effective in utilizing the local environment. If this environment were homogenous and stable, such knowledge might be acquired rather easily. The fact remains that it is not a simple place to understand and to degrade the information base upon which local decisions are made seriously impairs local adaptive abilities. Information which has taken millenia to build up can be lost in a single generation.

Tourism is another lowland influence which can disrupt the highland adaptations. The increased resource and human support necessary to house, feed, and care for tourists in a manner which meets both their pleasure and comfort levels puts additional strain on limited resources. It also exposes group members to a value system which clashes abruptly with their own and tends to undermine it. Community members experience a continuous flow of lowlanders who come laden with material possessions and who can buy services but seem not to engage in any productive activity. Although highland

communities obviously do adjust to tourism, they do so by
substantially altering their values and goals to the extent
that it is difficult to return the former adaptive pattern.
This would not be serious, except that popular areas for
tourism seem to shift. Thus substantial changes in the
highland human and environmental system may be made to
accommodate lowland tourists, who after several decades may
no longer be interested in that area. Finally, the larger
the scale of tourism (million dollar hotels), the more lowland
corporations and governments can be expected to control local
conditions.

In response to increasing influence and competition,
highland resistance is rather ineffective. These are small
groups of people without great power, who are unsophisticated
as to ways of influencing government decisions. As roads and
airplanes facilitate the access of lowlanders to such groups,
past strategies which relied upon the remoteness of mountains
for defensive advantages and concealment are less effective.
If governments can be convinced that highland groups should
be maintained, it is they who will have to control deleterious
effects from outsiders, including their own.

Factors Contributing to Degradation

Human factors which appear to contribute to the degradation
of the highland human/environmental system are as follows:
an increased emphasis on opportunistic strategies rather than
those emphasizing community cooperation; changes from an
equalitarian to a hierarchial social organization; increasing
dependence upon capital-intensive, technological solutions
controlled by relatively few families; and an increase in
individual consumption norms. These four factors interact
and reinforce one another in affecting the adaptive pattern.

Reviewing these in a hypothetical manner from material
drawn largely from the Andes, the following changes would be
expected (though not necessarily in the order presented):

1) Exposure to value systems which characterize local
    conditions, traditions, and residents as inferior
    and lead to a breakdown of group values supporting
    the adaptive pattern. Inferiority is frequently
    expressed in terms of insufficient material possessions
    and lack of knowledge about the national culture.

2) Dissatisfaction permits a greater incidence of
    opportunistic behavior on the part of individuals
    and households in order to meet increased consumption
    norms.

3) Such behavior leads to more frequent challenging or disregard for community regulations. Attempts are increasingly made to manipulate cooperative networks and increase the utilization rate of local resources. Finally, dissatisfaction with local conditions allows one to justify the sale of resources to outsiders.

4) Formalization of cooperative networks to prevent cheating reduces their flexibility. The institution of wage labor further weakens the possibility of relying upon this network. It also allows laborers from outside, who have little concern about community obligations or the local environment to enter the community and influence decisions.

5) Capital-intensive technological solutions to environmental problems, when controlled by relatively few families, can give these families a considerable advantage over other households and substantially increase the rate of resource utilization.

6) When such technological solutions or assistance programs encourage resource specialization (emphasis on a single crop or herd animal, mining, tourism) they reduce the resiliency of the human system. This is especially maladaptive if the resource can be exhausted within several decades or if its price fluctuates in an unpredictable manner.

7) Growing consumption norms and increasing disparity in wealth and power within a community with an unchanging productive base leads to progressive alienation of households from essential resources. Out-migration of a portion of the community follows as social-psychological and biological well being is adversely affected.

The above factors contribute to the breakdown of an adaptive pattern which is based upon multiple resource strategies, prediction of local environmental conditions, and flexibility in responses. As high-resiliency responses are replaced with those more typical of high-stability systems, we would hope that changes would also produce more predictable environmental conditions. While this can sometimes be demonstrated for certain families in an agrarian-based community, a closer examination suggests that it may be accomplished at the expense of other community members and sometimes the local environment. For these members conditions become less predictable and flexibility and production is reduced. Ultimately many are forced to leave the human/

environmental system they have spent their lives in and search for one with a better quality of life.

Obviously as highland groups become more exposed to lowland influence changes will occur. Consumption norms will doubtlessly increase and it is inappropriate to expect or suggest that highlanders be satisfied with what they now have. Formal education will likewise be sought as a means of gaining opportunities and interacting with the national culture. Specific responses within the adaptive pattern will be replaced and new ones added (83). I do not intend to categorize certain changes as maladaptive but to suggest that, when these reach a certain level or appear in certain combinations or forms, they can seriously effect the adaptive pattern. Education, for instance, will be increasingly indispensable for members of highland communities, but it need not be antagonistic to local knowledge and values. Likewise technological solutions (e.g., inexpensive irrigation pipes), which are available to the general community and do not degrade the environment, can supplement existing solutions. Finally, rising consumption norms and significant inequalities in wealth and power can be influenced by governmental regulations on the types of goods available and wealth redistribution reforms.

To some readers it may seem that this paper has placed too much emphasis on the consequences of lowland intervention. Effects that this has had on the Andes, however, suggest that prolonged and systematic exploitation of a region can result in long lasting degradation of human systems. Griffin (84) provides an example of this in his examination of the effects of regional interdependence between the highlands and coast in Peru before the more recent reformist governments.

> It is quite clear that internal migration trade flows and capital movements have had *absolute negative effects* on the level of per capita consumption and the rate of growth of the poorer region {the highlands}.

> Internal migration has resulted in a transfer to the cities of the most ambitious and skilled rural workers. Precisely the most valuable human resources of the countryside are lost to the urban areas, where they may spend up to ten years searching for work in factories. The government's policy of linking the coast and the Sierra with highways, by facilitating the exodus, has only aggravated the problem.

Table 4. Peruvian highland trade with coast in millions of soles for 1959.
Modified from Griffin 1969, p. 318 (84).

Category	Exports	Imports
Industrial Products	473	671
Commerce	99	476
Services	--	233
Agricultural Products	3002	174
Minerals	459	198
Finance	30	2
Other	131	89
TOTAL	4194	1843
Export Surplus	--	2351
	4194	4194

Even more important than the transfer of labor to
the coast has been the transfer of capital. The
interregional trade figures are eloquent in this
respect (see Table 4).

(Table 4) indicates that the Sierra had a surplus
of over 50 percent in its trade with the coast. This
means that its *level of consumption was lower than
it would have been had there been no trade*. Orthodox
theory would lead one to think that the export surplus
would be compensated on capital account through the
accumulation of deposits (and other assets) in the
region's banks. These deposits would constitute
savings for the region and accelerate its rate of
growth.

The orthodox presumption, however, is the reverse of
the truth. The Sierra's agricultural exporters are
latifundistas who live mostly in Lima and deposit
their receipts in the capital's banks.

Table 4 suggests that the Peruvian highlands have a high
productive potential. This was certainly the case in preconquest times when a number of civilizations culminating with
the Inca arose in this region. Even in the colonial period
when attempts to resettle Andean groups and restructure
social relationships resulted in significant declines in the
population, one gets the impression that this was still a
very rich area. Today in small villages elaborate churches
stand as an indicator of past wealth. Their present condition
of disrepair and the poorly dressed village residents who
pass them by suggest that better times have gone before.
Reasons for this are explained only in small part by excessive
deterioration of the biotic environment; instead it is
necessary to examine changes in the social environment which
channel surpluses out of the highlands.

This pattern is common also in the Alps where most
communities have been controlled or influenced by external
forces: the Holy Roman Empire, royal dynasties, feudal
lords, and the national governments associated with the
French and Industrial revolutions (18). Likewise in the
Himalaya interaction with lowlanders has influenced substantial
changes in highland groups. Stiller (23, quoted in 22) notes
that highland trade was of critical importance both economically
and politically in the formative era of the Gorkha Conquest.
More recently the Chinese influence in Tibet has severed
trade routes into Nepal, which in turn is pursuing a national
policy of development and economic unification. According to
Messerschmidt (22), Nepali industries are manufacturing

commodities for consumption in the highlands, and new roads
are being opened up which link the capital with formerly
remote regions. As in the Andes, lowlanders (Hindu castes)
dominate trade, government, and education in Nepal and in
fact have become the dominant cultural group within the
country. As highland groups become more dependent upon
commercial trade, shifts in their value system and social
organization which approximate Hindu culture follow. Messerschmidt notes an increased emphasis on individual success in
societies where strong lineage and clan ties have been
emphasized. Also, Himalayan highland groups are shifting to
hierarchial systems emphasizing patron-client relationships
and caste ethic. Herding has become low priority in many
villages accompanying a readaptation to lower ecozones.
Increased commercialization has put pressure on households to
produce goods which bring high cash value (19). Although
Nepali highland groups have not experienced the extreme
exploitation seen in the Andes, definite consistencies in the
pattern of control over mountain resources are beginning to
emerge. The major difference between the two regions is that
the Himalayan environment appears to be more susceptible to
disruption because of steepness in slope. Thus deterioration
of the high Himalayan human/environmental system can potentially
proceed at a much faster and more irreversible rate than in
the Andes.

Conclusion

Throughout this paper an attempt has been made to
generalize about the adaptive responses of self-dependent
groups residing in the more tropical high mountains, and the
increasing influence which lowlanders are having on highland
human/environmental systems. Deductive generalizations
derived from theoretical ecology have suggested that under
unpredictable conditions, a response system emphasizing a
high degree of resiliency is expected. They also suggest
that competition between groups who differ considerably in
their exploitative abilities reduces the range of habitats
and resources the competitive subordinate group can use; this
effect is exaggerated in groups occupying more marginal
habitats.

Data derived mostly from the Andes and Himalaya largely
support these and other conclusions. Tropical and semitropical
mountain environments are unpredictable as well as heterogeneous
and fragile. The pattern of human adaptive responses is one
based on flexible, cooperative strategies which allow environmental opportunities and problems to be redistributed among
group members. In this way, critical variables of the human
system remain within their tolerance limits.

Growing competition from lowlanders in the future is expected to reduce the number of resources that highland groups have access to. In doing so, the range of environmental conditions to which they can effectively respond is diminished and their environment becomes progressively more unpredictable. At the same time these groups are being influenced by lowland institutions in ways which appear to undermine their values and adaptive responses. In some cases degradation of the highland adaptive pattern results from lowlanders trying to impose responses appropriate for more predictable environments on a high resiliency response system. While the intention may be to assist, the consequences frequently drive critical variables in the system toward their tolerance limits. When these limits are passed, the human system becomes degraded, that is, the number of humans the environment can support is substantially reduced. Formerly self-reliant individuals who are forced to leave highland systems generally migrate to lower altitudes and into various forms of institutional dependency. Here they compete with lowland residents and demand government services. Thus there is a definite cost which lowlanders must bear as highland human/environmental systems are disrupted.

What is clear when examining change in the high tropical mountains is that competition with lowlanders will increase in coming decades unless it is regulated by the government. Individual highland groups are not in a position to effectively resist exploitative abuse. Hence governments must control this if they are at all concerned about maintaining self-dependent agrarian based peoples in the high mountains. Likewise they need to better anticipate the detrimental effects of their own assistance programs. While it is difficult to foresee the outcome of restructuring a human system through planned change, some adaptive guidelines have been provided as to what highland groups are doing right. If accurate, these could provide the basis of a more systematic framework to be used in the assessment of both the impact of changes on tropical high mountain systems and the supplementation of existing adaptive responses. At least it may provide a basis for judging what changes would further degrade highland systems.

Finally and most importantly, any government action must make use of the vast bank of knowledge which still exists among highland peoples. If administrators are humble enough to listen to highland solutions and aspirations they will gain a perspective that the best of lowland advisors cannot recreate. This point is best made by the late Andean writer and poet José María Arguedas (1911-1969) in a poem entitled "A Call Upon Some Doctors." The poem, which appears below in

abridged form, was written originally in the Quechua language
and dedicated to John V. Murra whose ideas have been liberally
used in this paper.   "Doctors" in this context refer to
lowland decision makers.

> They say that we don't know anything
> and we are backwards,
> that our hearts do not match the times,
> and they'll exchange our heads for others, better ones
>
> They say that some doctors tell this about us,
> doctors who multiply in our own land,
> who grow fat here
> and get golden.
>
> But doctor, do you know of what my brains are made
> and of what is the flesh of my heart?
>
> Start your helicopter and climb here, if you can,
> Take out your binoculars, your best lenses, and look.
>
> Below 500 kind of flowers of as many different
>   potatoes grow
> on the balconies of the canyons unreached by your
>   eyes.
> They grow in the earth, mixed with night and gold
> silver and day,
> Those 500 flowers are my brains, my flesh.
>
> Come close to me, doctor, lift me into the cabin
>   of your helicopter, and
> I will toast you with a drink of 1000 flavors, the
>   life of
> 1000 different crops I grew in centuries
> from the foot of the snowfields to the forests of
>   the wild bears.
>
> No!
> Did I work for centuries of months and years, in
>   order
> that someone I do not know, and does not know me,
> misshape my face with clay and
> exhibit me deformed before my sons?
>
> We don't know what will happen, let death walk
>   toward us.
> Let these unknown people come, we will await them,
> For we are the sons of the father of all the
>   mountains,
> Sons of the father of all the rivers.

References

.. Unesco, Programme on Man and the Biosphere (MAB), Working Group on Project 6: Impact of Human Activities on Mountain and Tundra Ecosystems, Lillehammer, 20-23 November 1973, Final Report. *MAB Report* 14, Paris: 132 pp. (1974).

2. E.P. Eckholm, The deterioration of mountain environments. *Science*, 189: 764-770 (1975).

3. J.W. Cole, Cultural adaptation and sociocultural integration in mountain regions. Paper presented in the symposium. Social Aspects of Mountain Communities, Fourth World Congress for Rural Sociology, Torun, Poland (1976).

4. L.B. Slobodkin, Toward a predictive theory of evolution. *In* R.C. Lewontin (ed.), *Population Biology and Evolution*. Syracuse Univ. Press, Syracuse, New York, pp. 185-205 (1968).

5. L.B. Slobodkin and A. Rapoport, An optimal strategy of evolution. *Quart. Rev. Biol.*, 49: 181-200 (1974).

5. G. DeJong, Demography and research with high altitude populations. *Soc. Biol.*, 17: 114-119 (1970).

7. E.J. Clegg, G.A. Harrison, and P.T. Baker, The impact of high altitude on human populations. *Hum. Biol.*, 42: 486-518 (1970).

8. Geographisches Institut der Universität Bonn (ed.), Geo-ecology of the mountainous regions of the tropical Americas. Proceedings of the Unesco Mexico Symposium, August 1-3, 1966. *Collquium Geographicum*, 9 (1968).

9. C. Troll, *Geoecology of the High-Mountain Regions of Eurasia*. Proceedings of the Symposium of the International Geographical Union Commission on High Altitude Geoecology, November 20-22, 1969, Mainz. Erdwissenschaftliche Forschung, Wiesbaden IV: 300 pp. + 25 plates (1972).

10. H.E. Wright, Jr. and W.H. Osburn (eds.), *Arctic and Alpine Environments*. Indiana University Press, Bloomington: 308 pp. (1967).

.. Unesco, Programme on Man and the Biosphere (MAB), Expert Panel on Project 6: Impact of Human Activities on Mountain Ecosystems, Salzburg, 29 January-4 February 1973, Final Report. *MAB Report* 8, Unesco, Paris: 69 pp. (1973).

12. J.D. Ives and K.A. Salzberg (eds.), Proceedings of the
     Symposium of the International Geographical Union
     Commission on High Altitude Geoecology. *Arct. Alp. Res.*,
     5(3, pt. 2): A1-A199 (1973).

13. J.D. Ives and R.G. Barry (eds.), *Arctic and Alpine
     Environments*. Methuen, London: 999 pp. (1974).

14. R.B. Mazess, Human adaptation to high altitude. *In*
     A. Damon (ed.), *Physiological Anthropology*. Oxford
     University Press, New York, pp. 167-209 (1975).

15. A.R. Frisancho, Functional adaptation to high altitude
     hypoxia. *Science*, 187: 313-319 (1975).

16. P.T. Baker and M.A. Little (eds.), *Man in the Andes: A
     Multidisciplinary Study of High Altitude Quechua*. Dowden,
     Hutchinson and Ross, Stroudsburg, Pa.: 482 pp. (1976).

17. P.T. Baker (ed.), *The Biology of High Altitude Peoples*.
     Cambridge University Press, Cambridge (in press) (1978).

18. R.E. Rhoades and S.I. Thompson, Adaptive strategies in
     alpine environments: beyond ecological particularism.
     Amer. Ethnol., 2: 535-551 (1975).

19. S.B. Brush, Cultural adaptations to mountain ecosystems:
     introduction. *Hum. Ecol.*, 4: 125-133 (1976).

20. I.G. Pawson and C. Jest, The high altitude areas of the
     world and their cultures. *In* P.T. Baker (ed.), *The
     Biology of High Altitude Peoples*. Cambridge University
     Press, Cambridge (in press) (1978).

21. R. Levins, *Evolution in Changing Environments*. Princeton
     University Press, Princeton, New Jersey: 120 pp. (1968).

22. D.A. Messerschmidt, Ecological change and adaptation
     among the Gurungs of the Nepal Himalaya. *Hum. Ecol.*, 4:
     167-185 (1976).

23. L. Stiller, S.J., *The Rise of the House of Gorka: A
     Study in the Unification of Nepal, 1768-1816*. Manjusri,
     Bibliotheca Himalayica, New Delhi (1973).

24. D.E. Shea, A defense of small population estimates for
     the Central Andes in 1520. *In* W.M. Denevan (ed.), *The
     Native Population of the Americas in 1492*. University
     of Wisconsin Press, Madison, pp. 157-180 (1976).

25. R. Peattie, *Mountain Geography: A Critique and Field Study*. Harvard University Press, Cambridge (1936).

26. C. Troll (ed.), *Geo-ecology of the Mountainous Regions of the Tropical Americas*. Ferd. Dümmlers Verlag, Bonn: 123 pp. (1968).

27. S. Hastenrath, Certain aspects of the three-dimensional distribution of climate and vegetation belts in the mountains of Central America and southern Mexico. *In* C. Troll (ed.), *Geo-ecology of Mountainous Regions in the Tropical Americas*. Ferd. Dümmlers Verlag, Bonn, pp. 122-130 (1968).

28. G. Budowski, La influencia humana en la vegetación natural de montañas tropicales Americas. *In* C. Troll (ed.), *Geo-ecology of Mountainous Regions of the Tropical Americas*. Ferd. Dümmlers Verlag, Bonn, pp. 157-162 (1968).

29. R.B. Thomas and B.P. Winterhalder, Physical and biotic environment of southern highland Peru. *In* P.T. Baker and M.A. Little (eds.), *Man in the Andes: A Multidisciplinary Study of High Altitude Quechua*. Dowden, Hutchinson and Ross, Stroudsburg, Pa., pp. 21-59 (1976).

30. B.P. Winterhalder and R.B. Thomas, *Geoecology of Southern Highland Peru: A Perspective on Human Adaptation*. Occ. Pap. No. 27, Institute of Arctic and Alpine Research, University of Colorado, Boulder, Colorado (in press).

31. G. Mann, Ökosysteme Sudamerikas. *In* E.J. Fittkau *et al.* (eds.), *Biogeography and Ecology in South America*. Dr. W. Junk, The Hague, pp. 171-229 (1968).

32. J. Tosi, *Zonas de Vida Natural en el Perú*. Instituto Inter-Americano de Ciencias Agricola de la DEA: Zona Andian, Boletin Technico No. 5, Lima (1960).

33. H.H. Shugart, R.B. Thomas, A.P. Vayda, D.L. Loucks, and D. Reichle, Viewpoints on Energy Flow in Ecosystems. Workshop Volume on Energy Flow and Human Adaptation, 1975, Gainesville, Florida, The Institute of Ecology (in preparation).

34. R. Margalef, *Perspectives in Ecological Theory*. University of Chicago Press, Chicago: 111 pp. (1968).

35. G.I. Child and H.H. Shugart, Frequency response analysis of magnesium cycling in a tropical forest ecosystem. *In* B.C. Patten (ed.), *Systems Analysis and Stimulation in Ecology*, Vol. II. Academic Press, pp. 103-134 (1972).

36. C.S. Holling, Resilience and stability of ecological systems. *Ann. Rev. Ecol. Syst.*, 4: 1-23 (1973).

37. R.V. O'Neill, W.F. Harris, D.E. Reichle, and B.S. Ausmus, A theoretical basis for ecosystem analysis with particular reference to element cycling. *In: SREL Symposium on Mineral Cycling in the Southeastern U.S.*, 1975 AEC-CONF (in preparation).

38. W.D. Billings, Arctic and alpine vegetations: similarities, differences and susceptibility to disturbance. *BioScience*, 23: 697-704 (1973).

39. N.W. Simmonds, The grain chenopods of the tropical American highlands. *Econ. Bot.*, 19: 223-235 (1965).

40. D. Ugent, The potato. *Science*, 170: 1161-1166 (1970).

41. M. Cardenas, *Manuel Plantas Economicas de Bolivia.* Impr lethus, Cochabamba, Bolivia (1969).

42. O.P. Pearson and C.P. Ralph, The diversity and abundance of vertebrates along an altitudinal gradient in Peru (in preparation).

43. C. Koford, The vicuña and the puna. *Ecol. Monogr.*, 27(2): 153-219 (1957).

44. R.W. Bullard, Vertebrates at altitudes. *In* M.K. Yousef, S.M. Horvath, and R.W. Bullard (eds.), *Physiological Adaptations: Desert and Mountains.* Academic Press, New York, pp. 209-225 (1972).

45. G.E. Folk, Jr., *Textbook of Environmental Physiology.* Lea and Febiger, Philadelphia, 2nd ed.: 465 pp. (1974).

46. P.T. Baker, Adaptation problems in Andean human populations. *In* F.M. Salzano (ed.), *The Ongoing Evolution of Latin American Populations.* C.C. Thomas, Springfield, Ill., pp. 475-507 (1971).

47. J.W. Eaton, T.D. Skelton, and E. Berger, Survival at extreme altitude: protective effect of increased hemoglobin-oxygen affinity. *Science*, 183: 743-744 (1974).

48. Z. Turek, F. Kreuzer, and L.J. Hoofd, Advantage or disadvantage of a decrease of blood oxygen affinity for tissue oxygen supply at hypoxia: a theoretical study comparing man and rat. *Pflugers Arch.*, 342: 185-197 (1973)

49. P.R. Morrison, Wild animals at high altitudes. *Symp. Zool. Soc. London*, 13: 49-55 (1964).

50. R.M. Garruto, Hematology. *In* P.T. Baker and M.A. Little (eds.), *Man in the Andes: A Multidisciplinary Study of High Altitude Quechua*. Dowden, Hutchinson and Ross, Stroudsburg, Pa., pp. 261-282 (1976).

51. L.L. Cavalli-Sforza, Genetic drift for blood groups. *In* E. Goldschmidt (ed.), *The Genetics of Migrant and Isolate Populations*. Williams and Wilkins, Baltimore, pp. 34-39 (1963).

52. N.E. Morton, N. Yasuda, C. Miki, and S. Yee, Population structure of the ABO blood groups in Switzerland. *Amer. J. Hum. Genet.*, 20: 420-429 (1968).

53. R. Cruz-Coke, Genetic characteristics of high altitude populations in Chile. WHO/PAHO/IBP, Meeting of Investigators on Population Biology of Altitude. Pan American Health Organization, Washington, D.C. (1968).

54. M.J. Fregly, Comments on cross-adaptation. *In* D.H.K. Lee and D. Minard (eds.), *Physiology, Environment and Man*. Academic Press, New York, pp. 170-176 (1970).

55. M.A. Little and P.T. Baker, Environmental adaptations and perspectives. *In* P.T. Baker and M.A. Little (eds.), *Man in the Andes: A Multidisciplinary Study of High Altitude Quechua*. Dowden, Hutchinson and Ross, Stroudsburg, Pa., pp. 405-428 (1976).

56. R.B. Thomas, El tamaño pequeño del cuerpo como forma de adaptación de una población Quechua a la altura. *Actas y Memorias del XXXIX Congreso Internacional de Americanistas*, Lima, Peru, Vol. 1, pp. 183-191 (1970).

57. A. Hurtado, Animals in high altitudes: resident man. *In* D.B. Dill, E.F. Adolph, and C.C. Wilbur (eds.), *Handbook of Physiology, Sec. 4, Adaptation to the Environment*. Amer. Physiol. Soc., Washington, D.C., pp. 843-860 (1964).

58. M.C. Monge, *Acclimatization in the Andes*. Johns Hopkins Press, Baltimore: 130 pp. (1948).

59. C. Hoff, P.T. Baker, J. Hass, R. Spector, and R. Garruto, Variaciones altitudinales en el crecimiento y desarrollo fisico del Quechua Peruano. *Revista del Instituto Boliviano de Biologia de la Altura*, 4(4): 5-20 (1972).

60. G. Murpurgo, P. Battaglia, L. Bernini, A.M. Paolucci, and G. Modiano, Higher Bohr effect in the Indian natives of Peruvian highlands as compared with the Europeans. *Nature*, 227: 387-388 (1970).

61. M.A. Little, R.B. Thomas, R.B. Mazess, and P.T. Baker, Population differences and developmental changes in extremity temperature responses to cold among Andean Indians. *Hum. Biol.*, 43: 70-91 (1971).

62. A.R. Frisancho, Growth and morphology at high altitude. *In* P.T. Baker and M.A. Little (eds.), *Man in the Andes: A Multidisciplinary Study of High Altitude Quechua*. Dowden, Hutchinson and Ross, Stroudsburg, Pa., pp. 180-207 (1976).

63. A.R. Frisancho, C. Martinez, T. Velásquez, J. Sanchez, and H. Montoye, Influence of developmental adaptation on aerobic capacity at high altitude. *J. Appl. Physiol.*, 34: 176-180 (1973).

64. R. McC. Netting, Of men and meadows: strategies of alpine land use. *Anthropol. Quart.*, 45: 132-144 (1972).

65. J.V. Murra, El "control vertical" de un maximo de pisas ecologicas en la economia de las sociedades Andinas. *In: Vista de la Provincia de Leon de Huánuco (1562), Iñigo Ortiz de Zúñiga, Vistador*, Vol. II, Universidad Hermillo Valdizan, Huánuco, Peru, pp. 429-476 (1972).

66. E. Beals, Vegetational change along altitudinal gradients. *Science*, 165: 981-985 (1969).

67. B. Winterhalder, personal communication (1977).

68. H. Kummer, *Primate Societies*. Aldine, Chicago: 160 pp. (1971).

69. T. Hagen, C. von Fürer-Haimendorf, and E. Schneider, *Mount Everest: Formation, Population and Exploration of the Everest Region*. Oxford University Press, London (1963).

70. C. von Fürer-Haimendorf, *The Sherpas of Nepal: Buddhist Highlanders*. University of California Press, Berkeley (1964).

71. R. McC. Netting, What alpine peasants have in common: observations on communal tenure in a Swiss village. *Hum. Ecol.*, 4: 135-146 (1976).

72. P.T. Baker and J.S. Dutt, Demographic variables as measures of biological adaptation: a case study of high altitude populations. *In* G.A. Harrison and A.J. Boyce (eds.), *The Structure of Human Populations*. Clarendon Press, Oxford, pp. 352-378 (1972).

73. E.R. Wolf, Closed corporate peasant communities in Meso-America and central Java. *Southwest. J. Anthropol.*, 13: 1-18 (1957).

74. E.R. Wolf, *Peasants*. Prentice-Hall, Englewood Cliffs, New Jersey (1966).

75. S.B. Brush, Kinship and land use in a northern Sierra community. Paper presented at the symposium on "Andean Kinship and Marriage". Annual meeting of the American Anthropological Association, Toronto (1972).

76. G. Alberti and E. Mayer (eds.), *Reciprocidad e Intercambio en las Andes Peruanos*. Peru-Problema No. 13, Instituto de Estudios Peruanos, Lima (1974).

77. S.B. Brush, Man's use of an Andean ecosystem. *Hum. Ecol.*, 4: 147-166 (1976).

78. J. Gero, *Andean Labor Exchange*. unpubl. ms. (1978).

79. R.B. Thomas, Human adaptation to a high Andean energy flow system. *Occasional Papers in Anthropology* No. 7, Department of Anthropology, Pennsylvania State University, University Park, Pa.: 181 pp. (1973).

80. C. Morris, *Storage in Tawantinsuyu*. Unpublished Ph.D. dissertation, University of Chicago (1967).

81. J.V. Murra, La funcion del tejido en vario contextas sociales del estado inca. *Actas y Trabajos*, Segundo Congreso de Historia Nacional del Perú 1958, Lima, pp. 215-240 (1962).

82. J.W. Wilkie, *Measuring Land Reform*. UCLA Statistical Abstract of Latin America: Supplement Series No. 5. UCLA Latin American Center, University of California, Los Angeles: 165 pp. (1974).

83. S.H. Forman, The future value of the "verticality" concept: implications and possible applications in the Andes. Paper presented in the symposium "Organizacion Social y Complementaridad Economica en los Andes" at the XLII Congreso Internacional de Americanistas, Paris (1976).

84. K.B. Griffin, Reflections on Latin American Development. *In* C.T. Nisbet (ed.), *Latin American Problems in Economic Development*. Free Press, New York, pp. 313-333 (1969).